# WHISKEY ONE

# Whiskey One

ALLAN READ

ISBN-13: 9781234567890
Cover design by: Allan Read
First Printing, 2024

# ACKNOWLEDGEMENTS

I would like to thank my wife Debbie,
Your unwavering support, encouragement and belief in me has
been my greatest motivation.

Mike and Tom
This adventure started during our Thursday afternoon sessions.

Sean
A memory of mateship.

The men of the 5th/7th Battalion, Royal Australian Regiment.
Without those faded memories, this book wouldn't be.
'Once a soldier, always a soldier.'

# Contents

# PROLOGUE

Honoured citizens of Proxima,

I stand before you today, not as a dreamer of distant worlds, but as a voice of reality. A reality shaped by our struggles and resilience here on Proxima.

I have never seen Earth. It exists only in the stories handed down by our ancestors. They spoke of blue skies, green forests and endless oceans. A paradise that now seems more like a myth than a memory. For us, Earth is distant, almost intangible, a cradle of humanity reduced to a legend.

When the first settlers arrived, they carried Earth with them in their hearts. A land of outstanding achievements, a birthplace of civilisation, art and history. But Earth also carried the weight of its mistakes. Mistakes that led to its demise, crushed by environmental ruin. The story is embedded in our collective memory, a stark reminder of what was lost and that we must never forget.

Here on Proxima, we walk a different path, one where survival is not a given but a daily effort. Our planet, though in the habitable zone, is not forgiving. Intense solar flares strip our atmosphere, leaving the surface barren and deadly. We live in domed cities, hidden beneath the ground, protected by walls of metal and rock. Outside, there is no air to breathe and no life to be found. The radiation alone could wipe us out.

Yet look at what we have built together. Our cities are marvels of human ingenuity. Self-sustaining ecosystems powered by nuclear and geothermal energy. We cultivate our food in hydroponic farms, recycle every drop of water and every breath of air. Within our domes, we recreate the lost echoes of Earth:

rivers, trees and farms. But we know it's only a simulation. The rain, the wind, the oceans belong to another world, a world that no longer exists.

And still, we have thrived. Despite these formidable challenges, Proxima has become our home. Here, we live in the delicate balance between perpetual day and eternal night, in that thin band where light and warmth make life possible. We've adapted, evolved and created a society that our Earth-born ancestors could scarcely have imagined.

We are no longer defined by the ideals of Earth. Our ways and our culture have diverged and rightly so. For generations, our ancestors clung to Earth's legacy, its values and its mistakes. But we have learnt that Earth's way is no longer our way. That world, beautiful though it once was, could not sustain itself. It collapsed, leaving behind warnings for us to heed.

Today, we are something else. We are Proximians. Through the trials of adversity, we mould our identity, sculpting it with our triumphs and defeats. We look not to Earth for our inspiration but to the stars, for it is there that our destiny awaits.

Let's focus on the future and leave behind what has been lost. We have made Proxima our own and in doing so, we have charted our own course. Rather than being bound by the shadows of Earth's past, let us be guided by the light of our own future.

Proxima is our present and it is amongst the stars that we will build our tomorrow.

Chapter 1

# WEE WAA - 2004
# (Earth Time)

Mike had an unwavering belief that he could defy the odds. Instant fortune could provide state-of-the-art farming equipment and a vacation for his wife. A long-awaited promise that remained unfulfilled for forty years. Without fail, he purchased his weekly lottery ticket with the same numbers and always grabbed the newspaper. This week, he included the monthly edition of the Rural Trading Post.

"Mornin' Mavis. The rains have eluded us once again," he remarked, highlighting the way the dark clouds in the distance had missed his farm. "At church this morning, Ol' Tommo said he had a drop, but not enough to trickle into his dams," he added.

"Yep!" Mavis replied as she scanned his purchases. Same conversations, different week.

He finished at the news agency and made his way back to his truck. While fastening his seatbelt, he licked his endlessly chapped lips. A burden brought on by the sweltering sun on this dusty, lifeless land.

"This is the one." He stated confidently, with a reaffirming nod to his wife. She rolled her eyes as she always did. The truck's starter motor gave an elderly moan and they headed back along the sun scorched road. They maintained their characteristic silence throughout the journey. The road disappeared into a dusty mirage, rippling and vanishing. Their eyes always scanning for a kangaroo or emu that didn't yet know the road rules.

"One day, this land will all dry up, you know," Jamima projected, with confidence in her prediction.

"I don't think so. It's like that El Nino thingy. The rains will come." Mike avoided reality with a touch of optimism.

Jamima's prediction was justified about both the land and Mike's lottery win, which was never to be. His bucket list was as common as millions of others who had the same futile dream.

Back at the farm, he lounged on his well-worn verandah chair. Mike fingered through the trading post, not intending to update his near-depleted machinery, but to dream. He fantasised about the fancy tractors that would find a space in his shed when he won the lottery. His fingers sensed it first, the middle page folded, heavier than the others.

Mike anxiously bit his lower lip. His eyes widened with excitement when he found an attached promotional scratchie. "What do we have here?" Curiosity filled his thoughts.

Ripping the scratchie along the dotted line, he separated it from the magazine. Becoming excited, he tapped the wooden verandah post for good luck. Mike scraped off the latex film with a coin from his pocket and enjoyed the same addictive rush he felt when the lotto numbers were drawn.

"You waste so much money on gambling. It would have been better if you donated at the church this morning." She mumbled, knowing it was going through one ear and blowing dust out the other.

With a beaming smile, he looked towards Jamima.

"You beauty!" His face glowed with excitement at winning first prize.

Ignoring him, she sipped her tea and swung her legs over the verandah's edge. Mike laughed at the absurdity of his win fall. He was now the proud owner of a distant, nebulous asteroid in an unreachable galaxy. Examining the ticket further made the situation more hilarious. Mike had the privilege of naming the asteroid. He chuckled at the irony of finally winning big, but the prize was worthless.

That evening, Mike worked on the endless tasks of repairing overburdened tractors with the necessary bush mechanics and duct tape. He ponded over the possible choices for a name. His disbelief in the idiocy of his win only resulted in humorous names. He even toyed with the idea of calling the asteroid Two Dogs, referring to his favourite joke. After exhausting his slate of laughable possibilities, it then became obvious. He named the asteroid in honour of his cold-hearted aboriginal mother, Jingai. After registering the name online, he threw the ticket into an old shoe box with other items that would create intriguing stories for his future grandchildren.

Since that day, Mike and countless ancestors have passed. Wee Waa has become a dry desert, no longer able to provide sustainable farming. Mother Earth now provides less of an embrace for her children. Mankind has ventured beyond the confines of Earth's atmosphere. After several generations, Mike's descendants began the process of dismantling the farm. Removing the old hardwood timber walls from the partially collapsed barn, they discovered a forgotten shoe box. On opening it, they uncovered what was inside, a worthless scratchie.

# Chapter 2

# PROXIMA

*Human existence has endured a further 286 years. With technical developments, a new evolution of homos has come to be.*

Kane recalled the history he learned as a young boy about Proxima. Humanity's second home and a new beginning on a planet light-years away. When mankind was young, the colonisation of other worlds existed only in the minds of stargazers. Now, Proxima stood as humanity's second civilisation after Earth, located in the Goldilocks zone of a nearby star. Despite its harsh, barren surface, settlers make life flourish beneath the planet's surface, with a labyrinth of subterranean cities.

Life in, not on Proxima is challenging. The surface is hostile with extreme temperatures, relentless windstorms, and deadly solar flares. The only natural resource capable of supporting life is a frozen, subterranean layer of ice. Thus, Proximians burrowed deep, creating cities that span miles underground, with soaring, cathedral-like spaces and interconnected tunnels. A lifestyle far different from Earths. Kane occasionally reflects on the resilience required to survive here, where people thrive despite the odds.

Amidst this rugged lifestyle, one of the burrowed spaces served as Mr Abbey's classroom, where he taught philosophy. A subject that seemed irrelevant to young Kane.

Kane remembers hearing about Proxima, humanity's second home, through the storytelling speeches woven into Mr Abbey's lessons. A new beginning on a planet light-years away. Once, the colonisation of other worlds existed only in the minds of stargazers. Now, Proxima stood as humanity's second civilisation after Earth, located in the Goldilocks zone of a nearby star. Despite its harsh, barren surface, settlers managed to make life flourish beneath it, with vast catacombs and underground cities.

Mr. Abbey began his lesson, his voice calm yet commanding as it filled the silent classroom. "Today, we're discussing Jingai," he said, pausing to let the name sink in. "A minuscule rock tucked away in the outskirts of a distant planet's orbit. An asteroid unknown to most people."

Kane looked around, noticing his classmates' blank faces. Nobody seemed interested in the idea of a rock floating somewhere in the universe, and Kane felt the same. Teenagers like him cared little about anything that didn't directly affect their lives.

"Why does it matter, sir? It's just a rock," one student remarked.

Mr. Abbey turned slowly to face the class with his calm gaze. "Yes, it's just a rock," he agreed, his voice unwavering. "But perhaps it's more than that." He paused, as if contemplating the importance of what he was about to say. "Jingai got its name from an ancient word, 'motherly embrace.' It's a name meant to evoke warmth and care, yet it's nothing but a desolate mass drifting in the void."

Kane felt a slight tug of curiosity. "Why would anyone name an asteroid after their mother?" he asked, a hint of skepticism in his voice.

"Who knows?" Mr. Abbey replied with a shrug. "Maybe it's more about the person who named it than the asteroid itself. Or perhaps it's about what the asteroid represents. A source of life hidden beneath an unassuming exterior."

Kane frowned, still skeptical. "So, why are we talking about it, then?"

Mr. Abbey's expression softened, a faint smile tracing his lips. "Because, Kane, even something as small and distant as Jingai has a role to play in the larger universe. For years, mining companies have been extracting valuable resources from Jingai's core, resources that are essential for sustaining life on Proxima and Earth. It quietly supports entire systems, yet most people don't even know it exists. Its impact is powerful, though unrecognised."

He turned to Kane with a knowing look. "And that brings me to you, Kane. Why is it that you want to join Proxima's security forces?"

Caught off-guard, Kane shifted uncomfortably. The question was direct, but Mr. Abbey's gaze held no judgment, only curiosity. "I guess..." Kane hesitated, searching for the right words. "I guess I want to do something that matters. Protect people. Make a difference."

Mr. Abbey nodded thoughtfully, as if he had expected this answer. "And do you think protecting people will make you feel like you matter? Being part of something bigger?"

Kane shrugged, glancing down at his desk. "Yeah. It's better than sitting here learning about rocks and space dust."

Mr. Abbey chuckled softly, but his eyes remained serious. "Maybe. But here's the thing, Kane. Significance doesn't always come from being seen or acknowledged. Sometimes, the things that make the biggest difference are hidden from view. Jingai's contributions may be quiet, but they're profound. Its resources sustain life without recognition or applause. It's already part of something bigger than itself."

Kane felt the weight of his teacher's words. He didn't fully understand them, but a flicker of realisation stirred within him. Even if others didn't always notice him, perhaps he could still make an impact, like Jingai, a quiet, unnoticed force in the universe.

As Kane grew older, he would rarely think of Mr. Abbey, yet the man's words became a subtle but steady influence in his life. His teachings; discipline, resilience, and the importance of critical thinking, had quietly woven themselves into the fabric of Kane's character. Now, as a member of Proxima's security forces, Kane often pondered his role in the grander scheme of things. Alone on night watch performing his security shift in the narrow corridors of the Centaurus, he wondered about his place in the universe.

The lessons he had learned from Mr. Abbey continued to echo in his mind, reminding him that sometimes, the most important work happened in the shadows. His job wasn't glamorous or celebrated, but it had purpose. His quiet efforts, though unacknowledged, contributed to the safety and survival of the people he protected. Just as Jingai's hidden resources had sustained countless lives across the stars.

Kane found solace in the idea that significance wasn't defined by recognition. Like Jingai, he had come to understand the power of quiet, unnoticed contributions. Whether anyone ever saw him, he knew that his work mattered. And in the stillness of space, above the surface of Jingai, Kane felt the truth of Mr. Abbey's words, carrying forward the legacy of that silent, unremarkable asteroid.

Chapter 3

# WHAT RAYNA WANTS, RAYNA GETS

Rayna was born within the damp underground city of Proxima, a place shaped by Earthen ancestry but distant from its roots. Her childhood was a warm, stable one within her family, who ran an engineering firm. It was here that Rayna honed her skills, rising to success as a designer renowned for her conceptual designs and recreations of replacement parts for heavy machinery. The demand for her talents kept her busy, but computer logic came so easily to her, she often felt her work was monotonous. The lack of challenge stirred something inside of her, a gnawing desire for excitement. Late at night in the workshop, her father, noticing her unease, leaned in to watch her draft.

"Another feasibility study?" he asked, a half-smile on his face.

Rayna sighed. "Seems like all I do these days. Nothing new. Nothing exciting. I feel like I could do this in my sleep."

Her father chuckled. "If you're that bored, maybe it's time to follow that dream you always speak of. Go and see more of the universe."

Rayna hesitated. "I would. I just don't know where to start."

The ache for something more had grown too strong to ignore. Rayna's father, seeing her restlessness, had used his connections to arrange a three-year teaching contract for Rayna. Now, she would share her skills with children aboard Jingai's BioStat. This journey wasn't just a job, it was her chance to break free from the routine that had confined her for so long.

As she prepared to board, her mother pulled her into a tight embrace, holding on as if to keep her grounded just a moment longer.

"Be safe out there, and don't let anyone underestimate you," she said, holding Rayna's face. "Remember, they'll always think they can tame you. Use that."

Rayna nodded, smiling. "When don't I?"

Rayna's first days aboard Jingai's BioStat were surreal. Surrounded by towering miners and engineers, her petite stature led many to mistake her as a student. A misconception she often found useful. Years of practice had taught her how to wield people's assumptions to her own benefit. Men usually saw her as either fragile or easily swayed and she knew just how to play into those beliefs. She'd add a dash of charm and quick wit to keep them off-balance. When Rayna wanted something, she had a way of getting it, with no one quite realising how.

But beneath her lively exterior was a serious mind, one shaped by Proxima's values, usually with a sense of duty to the future. Growing up, she'd been taught that they owed their progress to their forefathers. Proxima was a land without borders, so pride was rooted not in competition but in unity. A shared purpose of building a society together. As her mother once told her, "The only enemy on Proxima is stagnation."

Her academic years instilled in her a pride in their societies progress and the legacies of their forefathers. On Proxima, nationalistic ideals emphasised unity, not superiority over others. The people here took pride in the sacrifices and progress of their predecessors, fostering a sense of collective purpose that kept the society moving forward. But, as Rayna understood, personalities remain the same across planets with each community clinging to its own worldview. The distance between Proxima and Earth grew over the years, reflecting not only a physical separation but also a divergence in their societal beliefs. Proximians and Earthlings now struggled to bridge their differences, with social gaps widening and mutual understanding diminishing.

Walking between the desks in her new classroom when a student's question her about Proxima and its lack of churches.

She paused thoughtfully, "Well, in Proxima, we decided a long time ago that survival and progress were our main priorities. Religion..." she considered her beliefs before continuing, "...it didn't fit. Instead, we focus on Prolism. The idea that we all work together for the greater good."

Another student frowned. "So... no God?"

Rayna shook her head. "Not in the way that Earth thinks of it. We tie our beliefs to our purpose of making life better here and ensuring our survival. That's our version of faith."

She explains how minerals exposed on the surface, turned Proxima into a literal goldmine. Tidally locked to its star, one side of the planet faced perpetual sunlight. Rayna shared the stories about the other side of Proxima, a land of endless darkness and freezing temperatures where miners risked frostbitten limbs and sometimes their lives, for a chance at unimaginable wealth.

Life on Proxima was a contrast of wealth and hardship, much like the unpolished districts Rayna had known growing up. The planet was a treasure trove, an industrial haven and an ideal shipyard. Earth, once Proxima's lifeline, now depends on Proxima's resources to fuel its space ambitions. When she explained this relationship to her students, they often looked at her in disbelief.

"But Earth is supposed to be... well, better than Proxima, right?" one of them asked, perplexed.

Rayna shook her head. "Earth might have history, but Proxima has the resources. Times have changed. Now they rely on us for their expansion."

In confusion, the student asked, "So... who's more important?"

Rayna smiled, choosing her words carefully. "Neither. Earth and Proxima both need each other, even if they're slow to admit it. Without us, Earth's expansion would stall, and without Earth, Proxima wouldn't have the market share and influence it does now."

As the months passed, Rayna found a rhythm in her new life in the BioStat. She taught and learnt from her students. Most of her students had spent very little of their lives in planet. Rayna shared stories of Proxima's rugged landscape and the people who braved it daily. Now and then, when the hum of the space station grew quiet, Rayna would look out into the vastness of the stars and sense her relatives on Earth calling her.

One evening, as she stood by the observation portal, another young teacher approached her.

"Rayna, do you ever think about Earth?" she asked quietly, eyes fixed on the distant glow of stars.

Rayna smiled, a hint of wistfulness in her gaze. "Every day. One day, I'll make it there. Just to see what it's like." She paused, glancing over. "But for now, this..." she motioned towards the asteroid by the portal's edge, "...marks the start of my adventure."

Chapter 4

# JINGAI - 2090
# (Earth Time)

Most people call it BioStat, but officially it's the Biological Support Station, the only real lifeline on this desolate rock called Jingai. With no need for sprawling structures, the BioStat is a "trailer park" in space, or so everyone says, though few understand the term's origins. For Rayna, life here is a far cry from anything she's known. In an unnatural environment where life doesn't thrive as expected, strict rules, protocols, and a relentless focus on sustaining life are crucial.

Rayna leaned closer to the screen, eyes wide with excitement as she explained, "Dad, you wouldn't believe the BioStat! It's this massive wheel, nearly three hundred metres across, like one of those old-school motorbike tires, but it's alive. Picture a whole little world sealed up tight, full of plants, animals, all these smells, and the sound of running water. It is its own ecosystem." By speaking in an excited tone, she made her father think she was happy. "The locals call it a 'tin can,' but if someone opened this can, then poof! And everything inside could be lost."

She paused, letting the thoughts settle, then continued, "Inside, it's all life with rows and rows of vegetables growing in rotating farms, little patches of trees that actually smell like

Earth. There are racks of lettuce, carrots and herbs all threaded through pipes, fish tanks for aquaponics where fish dart around algae. It's wild, really a little oasis, but everything here has a job. Every leaf, every insect, every cycle is for survival."

She felt thrilled about learning new things. "They recycle the minerals extracted from the mine to create rich soil. Thek farmers then add bacteria and waste, breaking it down under lights that simulate some distant sun."

She took a quick breath. "They even have tiny animals and bugs for protein."

Her father just listened, knowing her excitement had taken hold.

Rayna gave him a knowing smile. "And us? We're the gardeners here...but we're also the ones who could ruin it all in a heartbeat if we're not careful. It's humbling, really, being part of something so intricate and fragile."

Unfortunately, the happiness she had shared with her father would soon seem far removed from her current situation.

Rayna knew she was part of this ecosystem, yet that fact left her unsettled. Since her arrival, she'd struggled to adjust. Her initial excitement had dissolved into quiet frustration. She had her lava lamp, a farewell gift from her best friend on Proxima which was a tiny memento that had once made her smile. But now, its cool blue light flickered on her walls, somehow more isolating than comforting. She'd lay awake, watching the wax blobs rise and fall, reminded of the pool on Proxima where she and her friend had held their breath underwater, marvelling at the shimmering bubbles floating to the surface. Sleep seemed more elusive with each passing night. Her mind tricked by that soft, glowing light that made her body ache to stay awake.

When her shift started, Rayna would drag herself to the small classroom, her exhaustion seeping into her lessons. Once full of passion and purpose, her classes became a monotony she struggled to shake. She'd push through, but at the end of class, she'd retreat to her quarters, her ceiling swimming with those eerie blue shadows, transforming her tiny room into an imaginary aquarium where she'd drift alongside her friend among imaginary fish. But reality would eventually sink back in. Hunger would bring her back to the reality of her cramped, windowless room on a spaceship attached to a barren asteroid.

Rayna had grown more withdrawn, and the community quietly respected her need for solitude, allowing her space. She ate alone, her movements mechanical, her mind distant. Afterward, she retreated to her room, lying on her back, eyes fixed on the ceiling bathed in blue light. She never even thought to switch off the lamp, simply staring upward, lost in thought.

As time passed, the scent of home seemed to drift closer, growing stronger and more vivid in her mind. Rising from her bed, she moved as if awake, though her eyes still held the haze of dreams. Rayna paused at a porthole and gazed at a dark space adorned with scattered stars. The idea of countless alternate realities added to her already overwhelming feelings of isolation and loneliness. She stared into nothingness, feeling her reality slip as the stars seemed to swirl, mimicking the soft undulations of her lava lamp. For a moment, she imagined herself floating, releasing her last breath, surrendering to the blue void as she would in the depths of a pool. The stars called to her with a bittersweet allure, filling her with a deep sense of longing and release.

Rayna's isolation drew her into a realm where reality, memory, and dream merged, the boundaries between them fading. She longed to shed the weight of her body, to dissolve and drift among the stars. Her grip on reality faded, replaced by an aching need to return to that place of floating freedom she shared with her friend. Chasing a lingering, imagined scent of home, she wandered through the BioStat, eventually finding herself at the docking port. Alarms rang as the command centre detected the breach, and security rushed to intercept her. But Rayna, lost in her memories, stared out the small porthole of the docking port's outer hatch. Her hand on the lever, her gaze fixed on her view, the only possibility of reaching such a vast distance. Rayna's chest felt tight. Her mind spun with a longing for the freedom. She felt her heart pounding. Then a faint voice crackled through the intercom, she realised the sound wasn't her heartbeat; it was someone who was knocking, loud and steady, on the inner hatch. Her gaze remained fixed on the portal, with the pull of open space overwhelming her senses. The stars seemed so close, calling her to drift free, to let go.

"I am alone. I miss my friend," she murmured, trying to appease the voice speaking to her through the intercom.

"You'll see her again," the voice reassured, calm and steady, but she barely heard it.

Her thoughts circled, each one a fragment of a memory from a different life. "I miss home! I miss Proxima! I miss the pool!"

"What's your name?" the calm voice asked.

Rayna blinked, finally turning her head, meeting the man's eyes on the screen. "Rayna," she whispered. There was something in his voice that made her pause, as if he truly grasped the tug-of-war playing out in her mind. But the weight of her memories returned, and she drifted back into them, mumbling to herself.

The man offered a small, reassuring smile. "Look at me, Rayna. I'm Kane. I can help you get home."

His words wrapped around her like an anchor, drawing her out of her haze. She opened her eyes, really seeing him this time for the first time. She looked around the airlock, a wave of disorientation washing over her. How had she ended up here?

"If you open that hatch, the void will pull you out," Kane's voice grounded her and she looked down, realising her fingers were still gripping the latch.

He spoke with steady resolve. "Rayna, listen to me. If you open that, you'll lose everything. You don't want that. The air will be sucked from your body." Kane knew she was now listening to his voice and kept reassuring her. "It's not what you are looking for. Let me help you."

Her hand fell away from the latch, his words cutting through the fog in her mind, jolting her back to reality. The weight of his calm voice steadied her. She turned, stepping back through the inner hatch to find Kane waiting with a medical team behind him.

Chapter 5

# THE WHEEL HOUSE

A small percentage of the BioStat is designated for living areas. Including quarters for approximately forty caretakers who maintain their environment. The mining staff and their families sleep in the remaining quarters. Having two BioStats on the shaft means there are one hundred and forty lives to feed and house.

Their home, the spinning shell above the mine, is where the community experiences a primarily mundane existence. Days unfold in a predictable, repetitive cycle. The damp mustiness of air conditioning mingles with the odour of perspiration. Everywhere they turned, the sights and smells of animal dung and the finicky waste systems were inescapable. This being nauseating for some and paradoxically comforting to others. The idea of open windows and unobstructed corridors evokes memories of a bygone era. In this compact environment, personal space is a prized commodity. The residents in the BioStat are innovative, repurposing items to enhance their quality of life. Amidst this backdrop, businesses flourish. Families gather and socialise and on this day and today, like every other day, the children pretend to study.

Individuals don't have the freedom to choose where they live and the colour of their walls. Those with the desire to venture off the grid can only walk as far as the pressurised hatches. Merely existing can weigh heavily on even the most resilient spirits. It's hard to imagine anyone enduring service on a Bio-Stat for more than a few years. Returning to your planet not only helps rehabilitate gravitational conditioning, but it also allows people to transition back into a normal society, maintaining mental stability.

Every few years, those in the mining community rotate, embarking on the long journey back to their home planet, Proxima. Most of the workers are descendants of the early pioneers that first burrowed into this subterranean site with their specialised robotic machines. As their robots toil away, the mining families and the neighbouring community carve out a gritty, yet prosperous existence. They remain blissfully unaware of the impending incident. It was a threat of such enormity that those on the ship would struggle to comprehend its scale.

After overcoming her struggles with anxiety and depression, Rayna found peace and contentment in her surroundings. Mental health issues are common, almost an expected in a life amongst spacefarers. Putting the experience into her past, she focused back on her role as one of the three teachers at Jingai. Despite getting used to her new home, the repetitive surroundings quickly became monotonous. Seeking more social connections, Rayna started spending her time at the recreation room in the outer BioStat.

One day, after wrapping up her class, Rayna changed out of her uniform and headed to the rec room to join the miners. As she stepped under the sign that read The Wheel House, she felt the weight of the miner's eyes on her, causing her to think, "They may as well piss in the wind." She thought. Ironically, she had never felt the instant panic of dodging warm urine that changes direction because of a strong gust of wind. It was just a saying she had inherited from her father, who had inherited it from his father, who had lived on Earth as a child.

The rec room had an inviting minimalistic ambiance, much different to the confined spaces of the rest of the BioStat, which can easily become overwhelming. The constant whiff of failing pumps and leaking motor lubricants, compounded by the hum and whine of machines, added to the feeling of being trapped. For those prone to claustrophobia, the experience of being inside any spacecraft can be deeply unsettling. The rec room was unique. It's a place with ample lighting, completely free from any pipes or clutter. A place to unwind with friends, where the crushing pressure of the cosmos fades away with each sip of mead. The classic depiction of miners with helmets and lights, chiselling at solid rock in search of rare minerals, is a thing of the past. Today, highly specialised technicians work as miners. They maintain and operate the robots that tirelessly extract every last mineral from the asteroid. Engineers wear claustrophobic suits while working in the depths of the tunnels. Thick clouds of dark organic materials form when mining asteroids, resulting in their suits getting clogged with dark soot. The engineers set up designated areas called 'havens' to remove the soot, enhancing visibility and increasing productivity.

The miners who delve into the asteroid's depths are much like any other labour force. As their shifts wind down, they eagerly anticipate the peaceful sanctuary that awaits them, where nothing out of the ordinary ever happens.

Rayna was well aware of the calibre of many of these miners, as they constantly showcased their skills to impress the new teacher. Knowing romance in such an isolated bubble most likely wouldn't end well, she just smiled and played along, commanded the hounds with impressive skill.

"You wouldn't know what to do with this, even if I let you get close." She saw through the miner's attempt to add another notch to his belt while showing off for his buddies.

"Give me a bit of your time and I will show you exactly what I can do." Davo grinned.

"A bit of my time?" she joked. "You're a sprinter!" His mates laughed at his failed attempt to bed her.

Davo chuckled despite the rejection and he offered his hand openly. Rayna reciprocated with a smile. Her response earned his admiration and they quickly became drinking mates. She found Davo's enthusiasm for electrical knowledge thrilling. His passion is to lead and innovate with advanced technologies. Rayna finished her recreation time and turned to her newest friend and bid farewell. "I could say goodbye to you, but in this can, I know I will see you around."

# Chapter 6

# BIBLE STUDIES

Rayna is the first in her family to proudly embrace her Proximian heritage. Her grandfather migrated to the planet as a small child. His parents raised him to uphold their family's prestigious titles and, above all, to continue growing the family's fortune. Contrary to Rayna's desire for adventure, her father's dream leans more towards the expansion of the family business dealing on Earth. Rayna's earliest memories involved her father reminiscing about the lavish lifestyle their ancestors had on Earth. The tremendous influence they wielded and the significantly better lives his parents had in those days.

Life aboard the BioStat had been a turbulent ride for Rayna, not at all what her father had envisioned. Initially, the novelty was exhilarating, but the constraints of the cramped living quarters had worn her down. Through a small portal, she once again found herself mesmerised, counting the interconnected containers passing by in an endless procession. They weaved away from the asteroid, carrying their precious minerals. In a video relay to her father, who had spent all his life on Proxima, she endeavoured to capture the essence of life on a BioStat.

"They call the long trail of containers that carry minerals away from the mine snakes." Rayna explained on one of her occasional calls to her father. Her gestures and animated tone brought to life the process of how they remotely shunt the carriage into the mine, where they are loaded to the brim with assorted minerals.

Then the conversation returned to Rayna answering with what her father needed to know. "We have one liner here. I think it's called the Centaurus. Sometimes, I spot its security teams around the BioStat. I would guess there are three or so." She remembers encountering one named Kane, who supported her through her depression but never discussed it with her father. Apart from the security teams, no one else on the crew leaves their ship. She knew her father would write this down to pass on to his superiors.

"Tell me more about the BioStat?" Her father asked.

She obliged and vividly portrayed what life is like on the Bio-Stat, explaining how the station is thoughtfully planned to ensure that each area serves a purpose. Whether the space is for storage, machinery, or the efficient arrangement of pipes, electrical units and thriving plants, their no space unused.

Engrossed in her own world inside her tin can, she would sometimes retell stories about her small community. "The farming units are much like those back home. But here they stack the crops in taller, tighter layers." Without taking a break, she described how, in one room alone, there can be anywhere from twenty to thirty layers spanning from the floor to the ceiling.

Rayna acknowledged the reality of life in space, "Space lacks true self-sustainability, but the technology utilised here effectively imitates the elements that support life as if we are on Earth."

She continued, "In every nook and cranny, plants of every imaginable type filled the space, from the farms to the gangways, sleeping quarters and even in the amenity rooms. It's like living in the heart of a farm facility." With a rush of enthusiasm in her voice, she conveyed how it was absolutely captivating. "The ecosystem here is home to various animals, including birds, bees and an assortment of insects."

This being one of those rare moments when she could offer knowledge to her father of experiences he hadn't encountered before. He usually controlled their conversations and now just listened to what she shared.

Her father then asked, "Do you wish you had worked on the farms during your teenage years?" He had suggested it back when she was studying, but their financial situation didn't deem it necessary.

"Yes, I guess I would have appreciated it."

With a fond smile, Dad reminisced, "I remember you always saying you'd love to see the open green fields of Earth one day."

"Yes, I really want to experience Earth's unique ecosystems; wind-swept mountains, damp forests, salty beaches. To feel the sun as water wash over the sand... I want to see it all." Thoughts of Earth's natural beauty filled Rayna's mind, a romanticised image of what it once was.

Back on Proxima, plants were common enough, part of her everyday environment. But here on the BioStat, the closeness and constant presence of vegetation felt different, more alive, almost essential. She didn't realise, but her fascination was an instinctive response to the fragility of her surroundings. A deep, subconscious recognition of how this carefully balanced ecosystem was key to her survival.

Her father questioned her focus. "How about your studies?"

Rayna, fully aware of her father's intention to discuss her Bible studies, replied, "Yes, Dad, I will continue to dedicate myself to my studies." She purposely responded with a brief reply, knowing that discussing the taboo subject of religion openly was not a wise idea.

Moving forward in the conversation, he asked, "So, after your initial thoughts, do you still find the miners as intimidating as you initially thought they would be?"

"The miners are nice people. Around a hundred and twenty people are working here, including their children. I would say that only about fifty are mine workers, though." At her father's request, she clarified the numbers of people as best as she could.

"I don't talk to them much. They mostly live in the lower Bio-Stat, but they seem nice." With an air of anticipation, Rayna responded to another of her father's questions, just as she had rehearsed the answers in her mind. Rayna wanted to reassure him that she appreciated his concern for her wellbeing and didn't want him to stress over her social life. Like many in their twenties, she had a tendency to see the best in people, assessing them based on their capacity to amuse her. Within her isolated tin can, the miners' imaginative thinking sparked excitement and amusement and she didn't want to miss anything.

Rayna quickly changed the topic and described her classroom and how exciting it was to meet her students and their families. Her father and her engineering skills got her this job, but she found more from passing on her knowledge in the classroom. Watching her students expressed excitement when finally understand something new always gave her pleasure.

"When they finally grasp a concept, it's like an imaginary light globe turns on above their head. It's so exciting! I love the satisfaction teaching gives." Suddenly, the video ended. Satellite relays gave low priority to personal conversations. A common struggle when trying to keep in contact with loved ones over vast distances.

The excitement of Rayna's new surroundings slowly dissipated and her world never changes. Her daily walks around the station have become increasingly claustrophobic. The same people, repeating the same tiresome stories and vulgar jokes. Complaints played on a loop, and some people turned out to be far different from what she'd expected. The optimism of her twenties was fading, yet Daddy's girl kept pushing forward, determined to keep a positive outlook on this once-in-a-lifetime adventure.

After enduring four more long shifts of cramming knowledge into mostly resistant minds, Rayna finally had a break from teaching. Amidst the small trees in the BioStat's farms, she found a quiet solace during her precious hours of leisure. Unlike her indifference towards farming back on Proxima, she discovered a sense of peace amidst the flora. Engaging in fruit picking, she relished the fragrant scents emanating from the farm. The fruit's freshness and vibrant colours are enhanced when the thermostats mimic a cool early morning. Starting at the top of the ladder and working her way down, gently twisting the fruit and filling her basket, sometimes glimpsing amazingly coloured insects. This newfound affinity provided a serene counterbalance to her usual bustling life. However, within this tranquil setting, interactions with others in this enclosed habitat were inevitable.

As Rayna helped pick peaches, she noticed Tom searching through the rows of trees. People knew Tom, the livestock manager, for his tendency to respond at length to casual greetings.

"How are you, Tom?" she reluctantly asked.

"Great, thanks, Rayna," Tom replied, "though my chickens might disagree, considering the amount of laying they've had to endure. But, we must eat."

Rayna harboured a particular loathing towards animal slaughter. She wasn't reluctant because she cared about the animals it was because her sheltered life protected her from the harsh process. She recognised the need for animal products in the BioStat's ecosystem, but found it unsettling that some farmers took pride in their craft. They not only enjoyed the act of slaughter, but eagerly shared every detail with others. This gave her an unsettling feeling in her stomach that she couldn't ignore.

Before Rayna could respond, Tom continued, "and once again, that white rabbit has disappeared. You haven't seen it, have you?"

Engrossed in his own thoughts, Tom didn't wait for a response. "I'd introduce him to the cooks if I had enough of his relatives to turn the soil."

Rayna smiled at the one-sided conversation as Tom continued his search, muttering about the life he loved as though it were a burden. The station bell rang, breaking the solitude of her surroundings. As the shift ended, she made her way to the mess to drop off her basket. Excitement filled her as she envisioned the fun evening with her miner friends at the tavern.

At the Wheel House, Rayna embraced the vibrant atmosphere of laughter and conversation. With ease, she merged into the crowd of individuals known for their rugged demeanour.

Davo had become a constant companion over drinks. He saw himself as a mate that she could rely on for support, where Rayna saw Davo as a big brother figure. Not the type of brother who offers protection, but rather the kind who is a reliable companion who has her back. Standing around their oval table, the systems officer accompanied them. With a couple of drinks in her, Rayna's curiosity made her want to understand her surroundings better. The officer began with a general overview, assuming Rayna had little knowledge of their world. This assumption was pretty spot-on because Rayna wasn't really into the subject, at least not entirely.

"The shaft that runs through the center of the upper BioStat and our BioStat continues down to the asteroid. All up, it's two hundred and eighty metres long," he explained.

"I'm riding the lift down the shaft for the first time during my next class." Rayna responded. "But I didn't know it was that long." She smirked. Her sexual connotations elicited some giggles.

The officer added, "I suggest organising a field trip down to the mine if the BioStats ever stops spinning. That'll be the only place on this rock with power and gravity!"

Rayna's persisted with the boring topic and asked, "Just out of curiosity, what happens if something serious goes wrong in the mine? The briefing mentioned the alarms and how we follow designated routes to the safety of our quarters. But what happens next?"

The officer reassured her, "Your quarters have emergency rations and it's self-contained with its own oxygen supply. You'll be fine there."

Davo added, "That's why we have periodic living quarter inspections, to ensure they meet the standards for survival."

"I see your point, but what happens if a catastrophic event in the mine threatens the BioStat? Do we just sit in our quarters and wait for death?"

The systems officer topped up Rayna's mug "No need to panic, the BioStat will eject and separate from the rock if we need it to."

Another miner lightened the conversation. "I would love to eject myself from the BioStat. The food here never changes and the coffee smells like sewage!"

Everyone laughed. Davo added, "I'd be out of here because I keep hearing the same boring questions from the same boring faces." Pointing his thumb at Rayna to his side.

The questions eventually faded, replaced by laughter that carried on until no one could drink any more. Rayna felt a growing unease about the impact her choices might have on these people. They were no longer just crew members, they had faces, personalities, and quirks. They were becoming friends, real people she cared about.

Chapter 7

# THE CENTAURUS

Many people compare liners, like the one currently registered to Jingai's mining company, with the old container ships that once navigated Earth's seas. Over time, these vessels have evolved through countless triumphs and setbacks into enormous, self-sustaining ships. They follow a blueprint that meets most space travel needs, with few designs deviating significantly from the original plans. Initially passenger vessels, they have transitioned to assert human supremacy across the expansive universe. These ships now serve the purpose of transporting cargo.

The Centaurus is a liner that's showing signs of age. Her primary duties include ensuring the provision of vital resources and safeguarding the mining facility on the odd shaped asteroid. Like all liners, she features an elongated shape designed to accommodate rows of modules and containers along her central spine. This spine accommodates three manned superstructures. The forward, mid and aft structures resemble three gigantic wheels spinning on its axis.

The superstructures are rotating cylinders that give the Centaurus a unique presence. The crew traverses between these three hubs along the central spine, navigating a ship that feels as alive and dynamic as its occupants. Thanks to the continuous rotation of the superstructures, everyone on board remains comfortably grounded, their feet planted firmly on the floor.

Each superstructure is a near-perfect reflection of the others, a testament to meticulous design. Each with a fully operational bridge, guaranteeing the maintenance of the ship's functionality and crew safety in the event of a disaster. The captain, embodying the spirit of vigilant command, rotates between these bridges, supervising the crews and ensuring that every part of the Centaurus runs like clockwork. This diligent oversight keeps the old, but steadfast liner in prime condition, ready to face the vast unpredictable expanse of space.

She meets the standards for liners but stands out where it truly matters. The mining company has generously invested in a top-notch crew to care for this aging ship. Continuous training and a commitment to excellence have instilled a deep sense of pride among her crew. With an effective leadership, the ship operates seamlessly. The crew maintains the ship's serviceability and attends to her every need. The captain of the Centaurus commands great respect from the crew, having rotated into her command many tours before. Leading with purpose and compassion, he prioritises his crew while honing his management skills. The crew respects his authority but understands that to a captain, the ship always comes first. They wouldn't expect it any other way. They know that without the Centaurus, they would be lifeless and adrift in an eternal darkness.

The superstructures resemble oversized mobile BioStats, equipped with life-sustaining biomes and all the essentials for survival. Essentially, it is a drifting micro-planet focused on carrying a mass of cargo. The structure's outer walls contain water compartments that undergo precise heating and cooling. The ship has escape pods and many tenders for transferring people and goods. Providing escape possibilities for all on board.

As mentioned before, each superstructure functions as its own entity, complete with its own crew. The captain and his team frequently rotate between command centres on all three superstructures and know most of the crew by name. The crews are organised into five designated categories: command, engineering, environmental, logistics and administration. Their exceptional skills directly result from continuous and rigorous training. The crew undergoes regular drills and simulations designed to keep them at the peak of their abilities. This ongoing commitment to professional development ensures they remain highly competent and prepared for any challenges they might face.

Red lights began flashing with the wailing of a siren, casting an urgent race throughout the ship's corridors.

The intercom wailed. *"For exercise, for exercise... Code three response... Fire! Fire! Fire! Fire in forward, mid and aft galley lifts."*

*"I say again... For exercise, for exercise... Code three response... Fire! Fire! Fire! Fire in forward, mid and aft galley lifts."*

Code Three grants off-duty crew clearance to stay in non-operational areas. All other crew members are required to scrabble to their stations. Part-time firefighters from the forward, mid and aft superstructures scrambled to be the first to bring the simulated situation under control. They all engage in a fast but coordinated effort to respond to the emergency scenario. Their swift actions highlight the effectiveness of their training, as they race against time to ensure the safety of the ship. The competition between superstructures is always competitive. Each wanting to be the best. Receiving recognition for their efforts is always a competitive bragging point.

*"This is the captain speaking... For exercise only... The fire in forward, mid and aft,"* His transmission paused as he checked his notes, *"forward, mid and aft galley lifts, has been extinguished."*

*"I say again... The fire in forward, mid and aft, galley lifts, has been extinguished. Well done aft on being the first to complete the simulation."*

The red lights and siren ceased. Life aboard the Centaurus resumed as though nothing had happened. In the aft superstructure galley, the high-fives were plentiful. It was another boost to the crew's morale and the captain doubled the victor's spoils.

The crew's competitive spirit knew no bounds, extending even to the most unexpected contests. Like who could grow the biggest tomato, who could manoeuvre a ship's tender with the finesse of a maestro and who could create a dance group with the best uncoordinated moves. It was always a spectacle to behold. There was even a competition to see which superstructure had the best captain impersonator, with the winner aware that their reward could just as easily become a punishment.

The captain announced for all to hear. *"Attention, attention. We congratulate Chef Laundess for winning the captain impersonation competition. For his significant efforts, I will pay for all his drinks at the next formal dinner."*

The kitchen erupted with cheers in celebration of Chef Laundess' triumph. He strutted proudly, proclaiming in the captain's authoritative tone that he harboured no fear of the captain's response being a less than favourable one.

The captain then followed with a second announcement. *"Attention, attention... Please note that Chef Laundess will be on duty during the upcoming formal dinner and regrettably won't be able to accept drinks from the captain. That is all."*

Laughter echoed throughout the Centaurus, knowing Laundo's victory celebration was short-lived. It was an obvious reminder that no one should impersonate the captain!

While the ships at speed, the bow shield provides protection against swift intergalactic debris and junk. Positioned forward-facing, the shield boasts millions of minuscule apertures that expel icy droplets with significant force to intercept small incoming objects. These droplets travel short distances before melting and slowing. The shield's rim recollects and channels the water into compartments for recycling in layered water compartments within the shield's interior. There are no absolute safety guarantees, but the system has showed reliability in mitigating encounters with smaller obstacles. However, the potential emergence of stray icebergs remains an unpredictable variable.

The crew retrieves supplies by accessing the containers arranged in the container bays between the three superstructures. When preparing the ship for a new transit, the crew clears out two expansive spaces, revealing the central spine in its entirety, a sight typically seen only during such refits. Before the Centaurus mission, the crew meticulously filled the container bays with rows of containers, stacking them until they reached the circumference of the superstructures. The bays don't just store supplies. They also accommodate portable BioStats, workstations, water tanks and emergency modules. Essentially, anything the ship or the mine requires.

Container spaces may also accommodate additional living quarters attached to individual gravitational rings equipped with life-support generators. To optimize the liner's capacity, they can stack low-priority containers outside the protective barrier of the bow's conical shield. An automated ejection system ensures the safety of those aboard the liner by activating if these containers sustain damage from atmospheric debris. The United Planetary Convention imposes hefty fines to offset the expenses associated with retrieving such space junk. However, affluent corporations are often willing to take risks and sacrifice a few containers in order to maximise their profits.

The Centaurus once carried some exotic cargo from an earlier time on Earth, including several containers of Egyptian artifacts, relayed via multiple liners from Earth. A Proximian buyer acquired these contents and is currently showcasing them on Proxima. The captain, who diligently oversaw every aspect of the loading and unloading process, notably felt stressed because of the significance of the delivery. Though he refrained from interfering, his presence emphasised the significance of the cargo. He was also mindful of the hierarchy above him and wanted to avoid any less favourable mishaps.

When a liner is not in transit, the repositioning of containers between the inner and outer sections of the ship can become a laborious task. In such instances, a lengthy cable suspends the containers in space. The cable serves the dual purpose of tethering the containers to the ship and supplies power, oxygen and water if necessary. If there is an emergency, the crew detaches the cable until they can retrieve the containers safely.

# Chapter 8

# SPECIAL EMERGENCY RESPONSE TEAM

SERT - Whiskey
Team Leader - Whiskey One
Ship's chief SERT Leader

The captain of the Centaurus is relentless with ensuring the safety of his crew. To fortify their security, he has employed three Special Emergency Response Teams (SERTs) ready to leap into action at a moment's notice. Each team comprises four well-trained members. With these daring responders at hand, Centaurus stands firm against adversity, ready to conquer any challenges that dare to threaten their missions.

SERT's undergo training to tackle a wide range of challenges, including counter-terrorism, piracy, civil unrest and breakdowns caused by cabin fever. These professionals operate under universal laws that delineate their boundaries, functioning as independent legal mercenaries. Unaffiliated with any government,

SERTs are hired by corporations to provide top-tier security services and asset protection.

The teams onboard the Centaurus are battle-hardened veterans. The ship's leadership has invested in these experienced professionals to ensure the utmost safety for their passengers, the crew and the companies' valuable mining assets. These veterans have undergone intense training and possess extensive experience in high-pressure situations. Their expertise and swift response capabilities are crucial in preventing and managing any crisis, making them an indispensable asset to the Centaurus' mission of maintaining security and support at Jingai's mine.

Kane enlisted in the Proximian security force right after turning eighteen. Now, several years later, Kane brings with him a wealth of expertise in military tactics, policing, security and anti-terrorism. The security force, although not a traditional military force, frequently creates the impression of being soldiers because of their utilisation of grey and brown camouflage, armour and specialised military equipment. Proxima has never had the need for an official army, as there are no defined borders or conventional adversaries. The disparities of wealth caused by the mining industry further deepen the divisions that have existed since colonisation. The economic divide has given rise to greed, which has led to organised social unrest and an escalation of terrorism.

As a child, Kane often lost himself in his own world of military strategies and basic tactics that he learned from history books and games. While his friends dreamed of space adventures and future monopolies, Kane immersed himself in "Chickenhawk" and "The Art of War." Strategy dominated his thoughts and became his greatest asset in situations requiring quick decisions. He knew he was never meant for a mundane job, using a pen or a spanner. Kane couldn't imagine spending his life doing the same thing day after day.

He grew up in a middle-class family with parents who encouraged their three children to enjoy life and pursue their dreams. Those, like Kane, who chose not to pursue college found their career options limited. His disinterest in academic pursuits only reinforced his resolve to follow his childhood passion, enlisting in the security forces. Now thirty-four, Kane is a seasoned expert in soldiering. The world of SERT is challenging, with short-lived careers due to high risks, long hours, and demanding tasks that limit personal life. Only the most dedicated and unconventional professionals endure and both traits are invaluable in a team. Kane isn't the loud or self-centred type who fits in easily at training facilities. His strength shows in the field, where he comes alive under orders to "secure and hold," thriving in the heat of action.

Known for prioritising duty over personal desires, Kane has built a reputation for problem-solving and reliability. Despite his career focus, he has picked up knowledge of liner systems through his own initiative. Kane's common sense and ability to get the job done makes him an indispensable asset to any commander.

Those who have befriended Kane often remain unaware of his first name. "Kane" rolls off the tongue smoothly and it is also common practice among police personnel to use surnames only. On the rare occasions when his first name appears on official documents, it's a surprise to everyone, including himself. The unfamiliarity of his given name never seems to stick and is quickly forgotten. Kane sometimes wonders if anyone aboard the Centaurus even knew his first name.

As with most military-style units, foul language is the norm. During intense moments or after a few drinks, Kane's team has given their respected leader various colourful and irreverent nicknames. He knows they're not being disrespectful, it's just that his approachable demeanour puts people at ease. His presence makes those around him feel comfortable.

Life aboard the ship is far calmer than policing on Proxima with no insurrections, few intense confrontations and much more time for reflection. Looking back, Kane treasures the memories of his service: the honour and loyalty shared amongst his teams. He often reflects on memories of his greatest strength, his ability to lead others through the most challenging situations. Whether navigating high-stakes missions or making tough calls under pressure, he's always been a steady presence his team could rely on.

With political and social issues declining back home, the demand for Kane's line of work has decreased. This shift gave him more opportunities to focus on his personal life rather than on protecting others. Instead of his usual well controlled existence, Kane found himself entangled in a string of unsuccessful relationships, which bred failure, doubt and confusion. Hesitancy and nightmares gradually replaced his once steadfast focus on control.

Kane's life became dominated by the dreariness of space travel, with frequent downtime and infrequent security issues. Despite his dedication, he questioned if he was losing his edge and whether he could still lead effectively when needed. These doubts weighed heavily on his mind as he navigated this unfamiliar territory.

Chapter 9

# ARTIFICIAL DAWN

K ane awoke, unharmed. The battle in his sleep had subsided, yet the remnants of his subconscious warfare linger. He sensed a gradual erosion of control over the fears that haunt his mind, a creeping vulnerability gnawing at his reality. His growing nightmares erupt with a crescendo of persistent tinnitus that greets him every morning. This high-pitched ringing, a permanent reminder of his constant exposure to deafening weaponry, serves as a cruel medal for his military service. Each awakening has become a weight of his past, relentlessly encroaching upon his present.

Kane found himself in a solitary dance with too much downtime. Rubbing the sleepiness from his eyes, he interacted with the artificial intelligence in his living quarters.

"Coms, how many full rotations until I can access my retirement funds?" The adrenaline rush of his earlier days, quelling riots and restoring order, now seemed a distant memory, replaced by the unexciting task of breaking up the occasional drunken scuffle.

Coms swiftly recognised Kane's voice amidst the silence and responded with its characteristic calmness. "You have fifty-four full rotations until retirement," Its female voice echoing faintly against the glossy walls.

With a groan, he started the arduous process of coaxing his heavy eyelids to remain open. The grasp of oversleeping was reluctant to release its hold. As he rubbed his eyes, the piercing buzz of the alarm he'd set to jolt him from sleep abruptly shattered the calm.

"Alarm off," Kane grumbled as the alarm fell silent. Then, with an irritated voice, he began his routine "Com's, light on."

Coms again broke the silence' "You have a message from Hulk."

Kane's brow furrowed in confusion. "Why the fuck did I choose a female voice?" he pondered, a twinge of annoyance creeping in at the reminder of his failures in relationships.

With a sigh, he commanded, "Play message."

Hulk's booming voice resonated. "Hey boss, once you've got your shit together, come join us in The Ring."

"Delete message and switch com's voice to male four," Kane mumbled.

"Message deleted," came the reply, now in a deep, authoritative male voice.

He recollected the nightmares that troubled his sleep, a frequent experience in his high-risk job. Kane battles not only the demons of those he defeated but also the haunting cries of the innocent lives lost. Seeking solace from his dreams, he reluctantly grasps for his medication. In his cramped quarters, barely larger than a walk-in wardrobe, he finds a semblance of personal space, yet neglects its tidiness because of the chaos within his mind.

Crew compartments offer more than a respite from the closed environments shared with fellow crew members. They also serve as self-contained emergency stations. Some living quarters can combine to accommodate partners or families. Kane, being single, enjoys the privilege of solitude in his modest, private living quarters. His only haven amidst the vastness of the ship.

Taking a moment, he rotated the plants in his room and gave his bonsai a few drops of water. Kane dressed in his duty uniform and shuffled zombie-like out of his quarters and down the corridor. As he walked past the open doors of his fellow crew members, Kane noticed the various trinkets decorating their walls. Travellers often cling to images of home, family, screen idols, and unattainable romantic fantasies. Having seen countless cabins since his teens and knows better than to make his too cozy The repetitive nature of their lives only deepens the sense of melancholy that comes with such familiarity. The faces in those images seem to grow more distant with time. Kane prefers to keep his focus on his team and their tasks, but allows himself one small indulgence in his cabin. He skims the library's database for novels until he finds one that captures his eye. Like most warriors, Kane believes in working hard and playing harder, leaving behind a trail of memorable episodes. Some are even more legendary than the rumours suggest.

As part of his regular morning routine, Kane stopped and momentarily gazed out of a porthole. It being the only porthole he passes within the labyrinth of corridors and bulkheads as he goes about his morning tasks. Portholes are a structural weakness that designers avoid. They are not for stargazing, but for the maintenance or operations of the external apparatus. Ship designers might never fully appreciate how crucial a tiny round window can be to countless weary minds.

Alone in his thoughts, memories swirled of home and old friends; and today, his previous lover. It was a bad day. He fought the yearning for companionship, someone to hold. Kane knew these thoughts were filling the void left by the lack of adventure in his current posting. So he synchronised his thoughts for the day ahead with the usually unnoticeable hum of the engines that generate the ship's artificial gravity. A passing snakes reflected a brief glint of light, momentarily interrupting his thoughts. Kane then noticed another crew member waiting, seeking a moment to savour the vast shimmering stars through the small round porthole.

He stepped aside, allowing the crew member to enjoy the expanse. "Any news about that grandson of yours?" he asked the engineer.

"Nah, not yet. Seems he's in no rush to join this universe." The engineer found it surprising that Kane remembered he was about to become a grandfather.

The usual stench of the amenities room reeked. Irritable bowel syndrome was common among the crew because of the temperamental ecosystems aboard the ship. The resulting odours left no illusion of sweetness. A harsh reminder that, in some ways, the adventure in his life was becoming crap. Were Kane's days of adventure truly behind him? Would he and his team eventually end up pot-bellied and bald? So, with an unusually defiant state of mind, Kane shaved, showered, and donned his soldier's face.

Kane gathered his headset and notepad before heading to the recreation room. This space was a rare oasis, free from pipes, plants, buttons, gauges or signs. A haven designed to maximise crew comfort. It buzzed with constant activity, serving as the liner's entertainment hub. Equipped to meet all needs, it offered an array of amenities: snacks, monitors and gaming stations. The ideal communal space to foster camaraderie among shipmates. It's fitted with areas for relaxation before shifts and the obligatory shots of the 'Eye Opener' after shifts. SERT referred to the rec room as The Ring. They spent a lot of time in The Ring, unfortunately it was often to break up fights or escort drunken crew members to their quarters.

Kane entered and scanned the room for his team. They weren't just colleagues; they were more like a family for whom he'd die for. The walls displayed screens adorned with rosters, ship movements and the familiar Proximian broadcasts, including a continuous stream of sports, news and reality shows. In the absence of a clear day-night cycle, The Ring always had crew members seeking stimulation. Its plush seating was prime real estate that could accommodate fifteen or more crew members lounging at any given time. For Kane and his team, it usually served as a chance to regroup and bond before their next assignment.

The low hum of conversation and the occasional clink of glasses filled the air. The scent of freshly brewed coffee added to the cozy atmosphere. Glancing up at the monitors, Kane lip-read the anchor, who began with a cheerful "Good Rotations everyone," before delving into the typically grim realities that accompanied the news. His eyes then found the familiar faces of his comrades, Pope and Hulk, seated on the lower level. Mai-huni, the fourth member of the team, was absent, which wasn't unusual. She often spent her time volunteering in the medical bay, driven by her obsession with honing her medical skills.

As he pulled on his boots, Pope fixed his eyes on a monitor. Hulk wore his usual expression of semi-consciousness, a constant state from the moment he woke until he went back to sleep. Kane noticed that both were unusually attentive to the news.

Hulk had been by Kane's side for many years and their bond had developed into something akin to a close brotherhood. Neither would openly admit to being inseparable, fearing mockery for showing vulnerability. Instead, they maintained a pattern of playful banter and witty remarks, a strategy to keep the mood light in their soldiering environment.

"Get off your ass, you oversized turd!" Kane announced his arrival.

Hulk turned away from the monitor, pausing the audio in his headset.

"I wouldn't be this big if I jumped around like you lot of wannabe ballerinas," he retorted, fully aware that his size was an advantage.

His chest patch read Sabbas Hullick, but everyone called him "Hulk." A nickname inspired by the massive bulk food containers carried by liners. Standing at an imposing height with a beefy build, he towered over everyone like a giant amongst men. His neck was almost nonexistent; swallowed by his expansive shoulders and chest. His limbs resembled unwieldy logs, always straining against the confines of his clothing. It seemed like he could never escape from spaces that were too small for his massive frame. An unmistakable odour clung to his body, a byproduct of his constant discomfort. His obsession with food was evident in every conversation, which inevitably circled back to his favourite topic; eating.

Kane and Hulk, seasoned soldiers with no grand career aspirations but plenty of common sense, thrived on the front lines of action. For them, that was living the dream.

"What are they on about?" Kane glanced at the monitor and lazily slumped into an empty seat, seemingly uninterested.

"Apparently some bloke bought a scratchie six hundred years ago and snagged an asteroid," Pope Shared, eyes glued to the screen. "His descendants found the ticket and now reckon they're owed royalties."

"No shit!" Kane responded, his go-to reaction to most things.

"Yeah, apparently, Earth is also claiming the asteroid belongs to them." Pope pulled on his other boot. He loved being the one with knowledge to share. "They've still got the scratchie their great-great-something-grandfather bought. Imagine that, finding an old ticket and becoming billionaires. Wish it was me."

Hulk brought him back to reality. "You wouldn't know what to do if you weren't shootin' shit and gettin' drunk with us."

"Oh, I'd still be getting pissed, just with someone prettier than you!" Pope shot back.

Hulk turned to Kane with a cynical glance. "They're saying there's trouble with our head shed at Proxima not accepting the ruling and refusing to pay up."

"Proxima?" Kane raised an eyebrow. Your own people making waves is always concerning.

"Yeah, the mine's in Proximian territory. And get this, it's the rock directly beneath us, Jingai," Pope added.

"Ah, damn!" Hulk suddenly exclaimed, spotting who took the next available seat.

Pope flashed a cheesy grin. "Looks like it's your turn to buy coffee," he teased. They had a wager on the department of the next person to take the closest seat and Kane's earlier arrival didn't count, since neither had bet on them being a SERT member.

"By the way, Kane," Hulk remembered, "the boss wants to see you in the head shed for a pow-wow at five bells."

Even though the news reports showed no signs of imminent danger, Kane sensed a shift was approaching. He contemplated various hypothetical scenarios and their potential consequences and considered what decisions command might make regarding security responses. With their current duties feeling as dull as ever, he found himself curious about what events could unfold, secretly hoping for some kind of change.

# Chapter 10

# THE DEGRASSE TYSON

Though expected to follow orders without question, security personnel often harbour a quiet disdain for the ship's executive. If the executive staff were to lose control, the soldiers would quickly rally under their own leadership, with one notable exception: the Centaurus' captain. Even a skipper of a lesser character than the captain would remain in command. No one is eager to risk another mutiny like the one on the Bounty.

The captain routinely had Kane summoned to the planning room for discussions and briefings. Normally, the security team received their exercises and assignments well in advance, allowing plenty of time for logistical preparation. But, as Kane entered the room, he sensed this wasn't just another routine SERT training. The presence of much of the ship's junior leadership underscored the importance of the meeting. Kane scanned the room, noting the tense, focused expressions on the executive team's faces, which confirmed his suspicion that something serious was on the agenda.

Kane found the captain deep in conversation with the head steward. "Be sure to convey my congratulations to chief steward aft. Her team's handling of the fire drill was commendable," the captain remarked. He then noticed Kane's arrival. "Good, you're here, Sargeant."

Lately, the skipper had grown thinner, his wiry frame seeming to shrink. His carefully combed-back hair was now thinning, revealing patches of shiny scalp that reflected the room's light. The chestnut of his hair had faded to a silvery grey, contrasting with the deep lines etched into his weathered face. Dense grey brows framed eyes that were still capable of a stern, commanding look.

Despite the physical changes, the skipper's presence remained formidable. Those eyes, sharp and penetrating, held their clarity, reflecting a mind as keen as ever. His memory was flawless, recalling details with a precision that defied his age. Conversations revealed a quick intellect and a biting wit, instantly dispelling any assumptions of frailty.

To outsiders, his slightly stooped posture and slow, deliberate movements might hint at the onset of decline. But those who knew him understood he was anything but diminished. His insight and judgment were as reliable as ever and his strategic thinking remained unmatched.

"Not a training exercise, sir?" Kane inquired.

"Far from it. This one's more up your alley," the captain replied. "Let's sit. The information officer is about to brief us."

The ship's executive team shuffled into the two rows of seats around the table.

Taking his place in the centre seat, the captain settled in and prepared to begin. Kane squeezed in with the others around the table, where space was tight and only standing was possible. As he found a spot against the wall, he couldn't help but feel a sense of tension. The murmurs subsided, and all eyes turned toward the captain, ready for the briefing to begin.

The lights dimmed, enhancing the clarity of the three-dimensional maps projected onto the table. On the wall, various images accompanied a copy of the intel officer's report. The formal briefing began with fundamental information designed to bring the less experienced staff up to speed. Most of those gathered around the table were familiar with their surroundings, having spent several rotations stationed above Jingai. However, the protocol mandated that all attendees receive accurate and current information during these briefings to ensure everyone was on the same page.

The intelligence officer began, "As per the United Planetary Convention, the rules on mining rights are clear. Any resources found within a two-light-year radius of a colonised planet fall under that planet's Sovereign Economic Zone, or SEZ." He redirected his audience's eyes with his pointer, "Now, while Jingai is technically outside of Proxima's SEZ boundaries, our home planet retains full authority to enforce laws against any unauthorised incursions. In short, Proxima holds the right to act against illegal mining operations."

He gestures to the monitor, highlighting Earth and Proxima. "Earth, the only other inhabited planet, also has its own designated mining zone. The convention prescribes all mining rights to both Proxima and Earth. Any unclaimed entities within ten light-years of either colonised planet automatically become part of that planet's economic zone."

The intel officer's assistant changed the projection to show the asteroid's trajectory. "Once an asteroid exits a planet's SEZ into universal space, the planet loses all sovereignty and rights that it has held. Though companies have the option to file a new claim through universal courts. As you know, mining is an essential and lucrative business and exporting to Earth usually provides great profits." Many in the room smirked.

He continued, "The claimant can hold a lease of no longer than 100 years. Companies keep the claim unless it is no longer safe or profitable. The asteroid's resources may deplete, or the cost of transferring greater distances may outweigh the needs and profits. Then choosing to withdraw is better than continuing to play pointless fees."

He then emphasised, "Our company legally has full rights to mine Jingai and deters any illegal incursions from Earth or unaligned companies."

Once gaining mining rights, companies send unmanned probes to drill, analyse and collect data. If the findings are positive, the possibility becomes lucrative for all involved. Establishment of the mine goes into full swing with profit in mind. As with all industries, they put processes into place to keep running cost at a minimum. They deplete as much of the asteroid as possible and get out. This is usually the case with many smaller mines. Surface construction is usually minimal after the initial tubes are constructed. The mining community often live in tubes underground; extracting resources and shipping goods to the buyer. Larger mines have a lasting presence, creating self-sustaining communities on the surface and sometimes even sprawling industrial complexes.

He continued, detailing the lead up to their current situation. When Jingai ventured into the vastness of universal space eleven months ago, it left Proxima's jurisdiction. At this stage, any entity can file a claim through the universal courts. The universal courts issue lease terms for either a hundred or three hundred years. It is typical for entities to make these claims well before the asteroid exits its SEZ. Once a company secures a mining claim within an exclusion zone, it typically keeps that claim until it relinquishes it. This is influenced by factors like resource depletion or the logistical challenges and costs associated with transporting goods over long distances.

The intel officer concluded by highlighting the obvious, "Mining is a crucial and highly profitable industry that drives economic growth and yields substantial profits that no one wants to relinquish."

"Proxima is currently dealing with administrative red tape related to this application," he indicated, pointing to the relevant document on the screen.

"What we know is that Earth has lodged a claim to Jingai, citing ownership supposedly gained through a raffle ticket from many years ago. They are arguing that SEZ laws are irrelevant because of ownership statutes. We expect this legal process to be lengthy and the courts may need considerable time to resolve it. While the outcome is uncertain, I estimate current laws will likely prevail. We're unsure about Earth's motivations in pursuing this issue, but their correspondence has taken on an unusually demanding and aggressive tone.

Kane remembered Hulk and Pope watching the scratchie incident unfold on the news. As the briefing progressed, he analysed the facts and assessed the odds of various outcomes. Drawing from a wealth of experience gained through countless briefings over his career, Kane knew that the most crucial information had yet to be disclosed.

The captain turned to the intel officer. "Jes, please continue and update us on the mine's background."

Jesse promptly instructed his assistant to change the images as he shuffled through his displays.

"Okay, people, a quick history lesson. Once the company secured the rights to explore this asteroid, they sent unmanned exploratory probes to drill cores, analyse and collect data. The findings revealed valuable metals, including gold, iron, manganese, platinum, rhodium and rhenium. What was once merely a possibility became profitable, and the establishment of the mine moved into full swing. As with all mines, they implemented processes to minimise costs, deplete as much of the asteroid as possible and then move on." He pointed to the asteroid on the screen. "This is the case with many smaller mines, like Jingai. Anyone who has been down there would have observed the lack of surface construction. The miners live in the two BioStats attached to the mine. There are twenty work-related tubed modules within the asteroid to maintain machinery and analyse minerals before shipping the goods to the customers."

"Thank you, Jes," the captain acknowledged with a brief nod. "Your turn, XO."

"Thank you, sir," responded the executive officer, patiently awaiting his turn. He remained seated, leaning on his elbows over the table, delivering his briefing with no additional aids.

"Privileges of being the ship's 2IC, I suppose," Kane thought. A sentiment likely shared by most in the room.

While the captain oversaw the Centaurus, the XO handled operations not directly related to the ship.

He began, "We have an incoming liner, the Degrasse Tyson." The room now visibly focused on the XO's words.

He continued, his tone steady. "The Tyson," He pointed to a monitor. "Is an Earthen liner that came out a generation after the Centaurus. Typically, it wouldn't draw much attention, but recent events and the cryptic transmissions from the Tyson have raised concerns.For those unaware, there's significant political and legal debate about this mine. As a result, we're taking extra precautions."

The XO added, "There won't be any support, as there are no other ships in this sector."

Kane remarked, "This is getting interesting!" feeling a surge of excitement. It wasn't the thrill of danger that thrilled him; rather, it was the sense of approaching a long-sought goal after intense training, with the prospect of success now within reach. As the XO continued, Kane's mind instinctively began sorting through potential scenarios and the logistics required, including the locations where the three teams would be stationed and the time needed for preparation.

"Under universal law, our company maintains governance over Jingai," the XO stated firmly. "Proxima has relayed their objectives. Therefore, our mission remains unchanged. The Centaurus will continue its mission to ensure the safety and security of the company's assets and the personnel working at the Jingai asteroid's mine.

Adhering to the standard format to communicate orders, he then launched into his situation report. These reports are always concise and delivered in a consistent format to prevent confusion.

"We have the liner Degrasse Tyson en route to our location, with an ETA of four…" he glanced at the monitor, "… and a half bells. The Tyson's intentions remain unknown, as does the identity of its cargo and personnel. The Centaurus will proceed to level two stations at the next bell following the conclusion of this meeting."

"Orders," the signal that all were to silence and listen.

"SERT Quebec is to secure the Centaurus at the next bell following these orders. Your team is to be ready for any circumstances that may arise. You are to remain in the mid structure until further notice." The orders are then repeated as per protocol.

The XO then asks, "Questions?"

"No, sir!" responded the team leader for Quebec, fully aware of his responsibilities.

The XO moved to the next. "Orders,"

SERT Papa is to proceed to the BioStat immediately and organise security requirements. Assess the requirements, then return to the Centaurus prior to four bells." He reiterated the order, then asked, "Questions?"

"Negative, sir!" Papa One's training easily covering such situations.

"Orders," Kane was ready to note his team's instructions.

The XO delivered his orders to Kane, instructing him to reconnoitre the Jingai asteroid mine site immediately with SERT Whiskey. Assess any security requirements, then return to the Centaurus prior to four bells. "Once back onboard, Whiskey One is to resume command of all Centaurus' security." The XO repeated his last order, then asked, "Questions?"

"The mine employees, sir?" inquired Kane.

"At this stage, I don't want to alarm the locals until we know more. If necessary, inform the workers to continue as normal. Say we are conducting routine security training."

"Copy that, sir," responded Kane. He remained attentive, but subconsciously he began planning.

The captain gives the XO a nod and takes over. "Logistics chief officer, deploy your tenders and retrieve the containers." The ship needs to be ready in case we need to make any unplanned movements.

The captain scans the table for any more comments. "It's currently one bell and thirty-two, ship time. I now conclude this meeting. May good fortune be with you and if things go sour, come home safe."

The planning room quickly emptied as the crew, well aware of their responsibilities, sprang into action. The SERT leaders immediately turn to their communication devices, sending out necessary orders and updates. Kane sends a warning message to his team: "Return to the squad room immediately. Mine security recon, light kit."

No matter what they are currently doing, the teams know it's time to get serious. Within moments, each member begins their preparations. Kit checks ensure all equipment is functional and accounted for. Whiskey performs weapon checks to confirm that they have loaded the firearms and engaged their safeties. Communication checks guaranteed all radios are operational and set to the correct frequency. They squeeze in quick toilet visits, aware that there might not be another chance for a while.

When Kane arrived, he found his team assembled and ready. Silence falls over the room as they await his briefing. Kane details the situation: "This is a routine visit to the tubes. There's no need for heavy equipment, just standard patrol gear. We'll drop off the supply box as we normally do in these situations, conduct our assessment of the mine and we should be back before lunch."

He scans the room, looking in the eye each team member to ensure they understand the plan. "Questions?" he asks.

Hulk couldn't resist. "Does anyone know what's for lunch?"

A few chuckles ripple through the team. This sortie wasn't much more than routine, which allowed for some relaxed humour. Kane shakes his head with a mix of amusement and exasperation. "Not now, you dickhead!" he retorts, prompting more laughter.

The team disperses to complete their preparations. Each team member goes through their mental checklist, ensuring nothing's forgotten. Then double-check their gear and make final adjustments to find some comfort. Kane takes a moment to reflect on the mission ahead. Despite the routine nature of the task, he knows that complacency is the enemy. He reviews the protocol in his mind, reinforcing the importance of vigilance and readiness. The team reconvenes and Kane gives the nod. They move out with precision, each member understanding their role and the importance of presentation to convey they are elite. As they head to the airlock, all the present crew stare. The status of SERT often makes a person stand taller and the usual pre-mission adrenaline builds.

They hear through their headsets that Papa One was also ready to depart for the BioStat. Whiskey boarded the shuttle that will take them to the mine below the BioStat. A quiet determination fills the atmosphere as they embark on the brief journey. Everyone's lost in their own thoughts, thinking about their role in the task ahead.

Security forces train to wield their weapons both left and right-handed. This allows the user more concealment whilst covering firing arcs in the tight areas around Proxima and on-board liners. The team arms themselves with only their SIPLATs (STANDARD ISSUE PISTOL - LASER AND TASER) since there aren't any threats expected. The SIPLAT is a single handheld weapon that is short and lightweight. Much like a handgun in appearance, but fires energy instead of projectiles. The operator receives the pistol grip and trigger assembly as one component. The grips recognition capability identifies the authorised user and their team members. While a ship's leadership has the power to countermand safety measures and remotely enable the weapon for the entire crew, it is not ideal to arm an individual on a spaceship. Especially if they are going through a psychotic episode.

With the pistol grip being the trigger, it's the upper segment of the SIPLAT that packs a punch. A laser unit, tipped with a taser and an interchangeable, lightweight, elongated methanol energy cell. The operator has the choice of four settings. The first option is safe mode, while the second is the muzzle taser, ideal for close quarters and crowd control. With just two clicks, the weapon activates its laser capabilities. In enclosed spaces with thin walls, the direct energy from the lasers is particularly effective. The operator can choose between two laser strengths: the first will incapacitate the target long enough to make an arrest, while the fourth click delivers a lethal dose. A target at a distance may survive a lethal strike, but in close quarters, death is the most likely result. The disadvantage of firing lethal is that the energy cell only holds enough energy for three shots.

SERT's patrol gear is notable for its distinctive features. Each member wears a versatile vest designed for quick access to a variety of specialised tools. AI capabilities in this vest aid in decision-making and offer enhanced protection against small arms fire. Additionally, it deploys countermeasures to absorb impacts from electronic weapons such as lasers. The AI integrates seamlessly with the user's weapon, offering real-time updates on remaining energy levels to keep the operator informed.The operator can magnetically attach their SIPLAT or any handheld weapon to their vest if they are carrying a rifle, ensuring easy access and quick release. When the operator's energy is depleted, they eject the spent energy cell. Returning it to their vest and recharge it by pressing it down onto a power pack secured around their waist. When all energy cells are depleted, the operator resorts to using a knife secured at the waist as the last line of defence.

# Chapter 11

# CENTAUR MEAD

Pope's physical presence may not immediately strike one as imposing upon first glance. He stands at an average height with a slim, almost boyish frame, looking like someone who's just hit his twenties rather than a seasoned individual who's seen over a decade of police service. His body showcases the wiry strength that comes from maintaining a healthy lifestyle, even though he often indulges in drink and revelry. His most striking feature, however, is his exceptionally smooth skin, a rarity for someone with his background. With no signs of age, as if time has neglected to leave an imprint upon his face, giving him an oddly youthful appearance.

Coming from a family with a rich tradition in law enforcement, spanning over several generations, he followed in their footsteps and joined SERT several years ago. Pope excelled in his training, relying on his quick physique and intelligence to navigate the challenges.

With unwavering self-assuredness, he emanates a confidence that surpasses his physical appearance. However, what he displays is not blatant arrogance, but a subtle sense of superiority that stems from his sharp intellect. Carrying himself with confidence, he's always ready to engage in discussions. Never shying away from asserting his opinions, no matter the subject. Often, he's right, which only bolsters his arrogance. Frequently bearing an angled smirk on his lips, as if privy to secrets the world is yet to discover. With an air of mystery surrounding him, he carries himself in a way that captures the attention of others and leaves them both fascinated and slightly unsettled.

Pope carries the air of someone in possession of inside information, always entering the room like a shadowy figure emerging from the depths of night. In contrast to his counterpart, Hulk, who tends to be more impulsive and reserved in expressing his opinions, Pope wields language with masterful precision. His thoughts flow smoothly and exactly, akin to poetry in motion. When he's not swearing, he speaks with deliberate grace, his carefully chosen words delivered with an air of refined eloquence.

Yet, like all complex individuals, Pope has many layers to his personality. One night of excessive drinking stripped away some of those layers, revealing a side of him that rarely sees the light of day. The mask of composure and self-assurance slipped, offering a glimpse of vulnerability and raw emotion. It serves as a reminder that even the most poised and confident individuals experience moments of fragility and inner turmoil.

Because of an unexpected vacancy, Kane had the responsibility of interviewing in order to fill a position on his team. Sadly, the team member lost his life while visiting an eco-farm with his new and unconventional girlfriend. He had embraced her love of the farms, relishing in the scent of some fresh plant life. He did not know that the local bees would be his unexpected adversaries. In an instant, a severe allergic reaction took hold, rendering him unable to speak his last words as his throat swelled rapidly.

So, an interview panel was established to find his replacement. Kane, along with two of his superiors, made up the interview panel. He also included a team member to provide valuable input from someone who would be working closely with the potential candidate. This time it was Pope, well, not the usual Pope. He was not too keen to be at the round of interviews because of a heavy night of centaur mead. A drink, named after the ship, created by fermenting a blend of locally sourced honey, fruits, spices and water together. Some say the name loosely takes inspiration from the strength of a mule's kick, resembling a Greek Centaur. Pope's ability to embrace both work and play with equal passion undoubtedly showcased SERT's renowned reputation.

The aspiring candidate passionately voiced his desire to join SERT, emphasising that it had always been his lifelong dream. Then, rather ill-advisedly, he suggested that his personality was the epitome of what the team needed. With that comment, the pounding pressure within Pope's head intensified, like a relentless mule kicking, shattering any remaining sense of professionalism in the room.

"Your personality, you say? It's never about you, or your fuckin' personality. You wanker!" Pope unleashed.

With his attention solely on the candidate, he ignored the throbbing pain in his head. "We are a team and your personality doesn't have room for our team!"

"And what you want," His voice raised. "What you want doesn't matter. No one gives a fuck about what you want!" The candidate could do nothing but receive Pope's rant.

"If your thoughts are solely focused on yourself, you will never have the privilege of wearing this badge. Pope points to the rectangular badge on his left upper arm. The team's badge comprises two triangles forming a rectangle, one triangle maroon, the other yellow with a wild pig's head in the centre.

The expressions in the room said it all. Some mouths curled into smirks, while others hung open in shock. Everyone focused on Pope, who had a sudden realisation that he should have kept his thoughts to himself. The unexpected turn in the interview became abundantly clear, leaving Kane to pick up the pieces. The candidate that caught Pope's attention clearly did not align with the team's values. Thankfully for the better, as Maihuni was the successful applicant.

Chapter 12

# SIX-THREE-CHARLIE

The shuttle underwent a thorough decontamination and the Centaurus crew loaded its cargo. In the intricate balance of ecosystems, germs present a perpetual threat, with the potential for catastrophic consequences if alien pathogens infiltrate the BioStat. To prevent such a disaster, the team adhered rigorously to sterilisation protocols before descending into the mine. The ship's computerised registrar logged the team's declarations with precision.

A synthetic female voice requested, "To proceed, state your identification number."

"2-3-3-4-7-0," Kane replied, then focused on the orange dot for facial recognition.

"Sergeant Kane, what is the purpose of your transit?"

"Security analysis of Jingai Mine," he answered.

"Sergeant Kane, have you exhibited any symptoms of illness during the last ten shift rotations?"

"No." He had long since learned to omit any mention of his sensitive stomach, a lifelong condition.

"Sergeant Kane, have you had any interactions with individuals who have experienced illness during the last five shift rotations?"

"No."

"Sergeant Kane, are you planning to transport any un-processed plant or animal products?"

The questions kept coming, one after another and Kane responded with five more "No's."

"Sergeant Kane, you are cleared to board."

The rest of the team completed the procedure and entered the tender for the short flight to Jingai. Tenders serve as the backbone of any liner, facilitating seamless operations as they effortlessly manoeuvre various types of containers in the gravity-free environment. These versatile craft serve as portable cockpits, attaching to containers, shuffling them around liners or travelling short distances. In this case, passenger transport, the pilot docked the tender on top and opened an access hatch to facilitate interaction with those on board.

The trip to the mine was routine, as the pilot didn't have to contend with gravity or a rapidly moving orbit. The bridge acknowledged the departure of the small tender and the pilot took control of the instruments after detaching from the Centaurus. Leaning forward in his seat, he glanced down through the access hatch to scan his passengers below.

"Ladies and gentlemen, welcome aboard Six-Three-Charlie," he announced, referring to his tender's call sign. "Please inform us if you need our dining service or a light refreshment," he joked.

Pope, with a grin, expressed his desire for a schooner of centaur mead, playing along as if it were an option.

Kane rested at the beginning of the transit, perhaps even drifting off for a nap. The lack of gravity removed the discomfort of his kit. Loosening his seatbelt, he shifted to the end of the bench seat, making himself comfortable in the corner. He daydreamed while watching the departure on the tender's monitor. The apparent stillness of the tender amidst the surrounding motion of the Centaurus spinning constantly awed him. Kane imagined himself standing on stones in a stream, watching the water flow past. An odd thought, as the only flowing streams he has seen were artificial. This altered reality swiftly faded, replaced by a peaceful, deep sleep.

"I'll tell ya what I reckon, Trashy," Pope opened the banter.

The pilot was eager to hear the humour coming from below.

"We need to request this bucket as our personal transport," Pope continued. "We could put a bar in your little cockpit and you could serve us drinks. All those flashy controls could be on autopilot, allowing you to do something worthwhile for a change. As it stands, you do fuck all up there."

The pilot responded, "Why should I dedicate my life to serving you worthless shits when I could transport more valuable cargo like alcohol, gourmet food and women, which I occasionally do?"

Hulk's ears perked up at the mention of food.

"And what's with the name Trashy?" the pilot asked.

"Ha ha ha, because you wish you hauled all the finer things in life, but we all know you just haul the trash." They found the friendly banter entertaining and laughed, using it to pass the time, unless it was directed at you.

Most security forces have standard lightweight patrol gear, but SERT stands apart with their specialist equipment, which puts them in a different league. Less policing, more military-style killing machines. Their uniforms feature brown and blue blocked grid patterns, ideal for the caverns of Proxima. Though the camouflage is less relevant in a ship's enclosed surroundings, the patterns bring a sense of familiarity to most. The fabric is a polyester weave bound with a conductive material designed to reduce currents from electroshock weapons. This protection has somewhat proven to render small arms ineffective, although the wearer's survivability is significantly reduced when exposed to high-current attacks or shots fired within close range.

"Has everyone checked the serviceability of their hemstats?" Maihuni asked. Hemstat applicators are injected into a wound, expanding like a sponge to stop the flow of blood and seal the injury. At the same time, they provide a drowsy but necessary pain relief.

"Sure have, Huni," Kane responded, using one of her few nicknames.

"Just remember, Kane, you're always welcome to taste this honey," she joked, knowing it would get a rise out of the others. He would never cross that line.

"What about you, Pope? How's yours?"

"My what?" he asked.

"Get with the program, Pope!" she scolded, like a mother reprimanding a child. "One of the most powerful tools in your arsenal is your ability to listen." She pointed to her ear. "You don't use this tool effectively. You try, but your mind often jumps ahead to your mouth and that one is overused and frequently out of its league."

Somehow, Pope listened and reflected on her apt choice of analogies.

"A skilled person listens uninhibited, ignoring other tools until they're needed. Got it?" He smirked. The others regularly ridiculed Pope for being too clever.

Hulk quickly added, "I'm good!" before she could come at him with the proverbial wooden spoon.

The banter eventually quieted down and the team rested until the pilot announced their impending arrival. The break in silence woke Kane, who felt he enjoyed the best sleep he'd had in a long time, attributing it to being surrounded by people he trusted.

Kane recapped the mission. "We'll stash the backup supplies first. Then we'll split into pairs and do a quick familiarisation of the tubes."

The team responded with slight nods, signalling their readiness.

"Make sure your medkits are in working order," Kane added.

"We've already checked them," Maihuni replied confidently.

Kane smiled, pleased with their preparedness.

As they waited for the docking process to complete, Kane filled the time by asking what everyone planned to do once their time with SERT was over.

Pope was the first to respond, his voice brimming with confidence. "Politics."

"No shit," Hulk muttered, having heard Pope's grand plans to fix Proxima's woes more times than he could count.

Kane chuckled, recalling how Pope's brash behaviour at Maihuni's job interview might fare in the cutthroat world of politics.

Pope turned to Hulk with a grin. "And you? Opening a pizzeria, maybe?"

"I haven't really thought about it," Hulk replied, then added with a shrug, "But I like the idea."

Pope then shifted his gaze to Kane. "What about you, Boss?"

"I'm a lifer," Kane said with a smirk. "I get my kicks by telling you lot what to do."

The three of them then turned their attention to Maihuni. The first three answers had been predictable, but she was the wildcard. Kane half-expected her to mention medical school or medivac duty, but her answer caught them all off guard.

"A mother," she said simply.

A heavy silence settled over the team as each of them tried to process the unexpected response. The idea seemed almost foreign in their line of work.

Finally, Pope broke the silence with a teasing grin. "Well, you've got three blokes here who can help you get that career started."

Laughter erupted, even from Maihuni, who found it amusing, fully aware that at least two of them didn't stand a chance.

Kane chimed in, "Why would you want that? You're already looking after three children here."

Before anyone could respond, the flashing green light cut the conversation short. Docking was almost complete.

Chapter 13

# SAFETY AND SURVIVAL

Beneath the craggy exterior of Jingai, the twenty-three subterranean work modules hummed with energy. Resembling metallic veins running through the rock, they occasionally breach the surface, connecting miners to the sparse atmosphere through a series of airlocks. Ingeniously engineered, each module features a stationary core encased in a rotating shell that spins perpetually around a protective radiation shield. These rotating shells act like the gears of an ancient timepiece, their synchronised revolutions creating artificial gravity, just enough to sustain human life in this alien environment. However, the relentless spin is a double-edged sword: prolonged exposure to the uneven pull of centrifugal force can trigger space adaptation syndrome. A condition similar to severe motion sickness, twisting the body's senses. Over time, the human mind adapts, but the creeping threat of radiation exposure remains, a silent shadow over those who dwell there.

The mining operations on Jingai stretch far beyond the subterranean modules. Robotic excavators burrow deep into the asteroid's core, extracting the precious minerals that fuel the colony's existence. From the modules, engineers tirelessly monitor and fine-tune the machinery, ensuring a steady stream of resources for distant markets. For a select few, this life offers the promise of wealth, but for most, it's merely survival, a relentless grind under the indifferent stars.

Amidst this sprawling industrial web, one room stands apart. Unlike the other spaces packed with tools and humming data processors, this chamber is about to be used as a temporary classroom. Twenty-three teenagers, the children of Jingai's miners, gather for their lessons. They are the next generation, raised within the cold embrace of the asteroid, their minds shaped by the hum of machinery and the distant dreams of Proxima. In this sterile environment, their imaginations flicker like faint stars, longing for a future far beyond the endless routines and monotony of mining life on Jingai.

Today, however, the students have the rare chance to venture beyond the confines of the BioStat, the central hub of their lives, for an educational excursion. Their teacher, Rayna, guides them down the narrow lift shaft, her voice cutting through the usual silence.

Despite being far from Proxima, Rayna still took pride in maintaining her appearance, especially when mingling with the mine's Proximian men. An attractive woman, Rayna cherished her long, ash-brown hair, finding comfort in the way it glides effortlessly across her skin. However, practicality often trumped vanity in the harsh environment of the asteroid. To meet the safety standards required for today's lesson, she pinned her hair into a low bun, a compromise between style and regulation. Carefully, she placed her snoopy cap over her head, tucking away any loose strands behind her ears. With her hair secured, she was ready to don her helmet for the day's lesson on weightlessness.

"It's always good to get out of our shoebox now and then," she jokes, attempting to spark some enthusiasm with her students.

The BioStat's central spine clings tenaciously to the asteroid's surface as they descend within its lift. A façade of indifference hid the excitement and nerves rippling through the students as they brace themselves for the near-zero gravity awaiting them in the tunnels below.

Their first stop is the preparation room, where miners gear up for their shifts. The air is heavy with the smell of machinery, a pungent mix of grease, metal and asteroid dust. The students wrinkle their noses in unison, some groaning in discomfort.

"Do we really have to be here?" one girl grumbles, clearly unhappy with the disruption of her routine.

"Safety and survival," Rayna responds, her voice unwavering. "We practice these procedures because they are crucial to our survival." The directness of her tone quashes any further complaints.

Yet, even as she speaks, Rayna's mind drifts to the unanswered questions that gnaw at her. How much longer can she endure this dull existence?

"Miss, I'm feeling strange," a student's voice pulls her abruptly back to the present.

"That's normal in the tubes," Rayna replies, seizing the moment to pivot into a lesson on space adaptation syndrome. The students listen closely, their discomfort fading as they focus on her words, absorbing the realities of the present dangers of life in space.

The lesson continues as they move to the Extravehicular Mobility Unit Station, the "suit-up" zone.

Here, the students carefully began to suit up, preparing for the transition into the airlock that leads to the tunnels.

Rayna's allocated suit was not a standard size. "All sweet things come in small packages!" were the reassuring words her father would follow with, when her lack of height was an issue. Anticipation buzzed among the students as they prepared for the near-weightless environment. Yet, the steady hum of the rotating modules and the metallic clank of a docking tender securing to a nearby airlock temper their excitement. The constant mechanical symphony reminds them of the delicate balance between survival and the vastness of space, reducing their enthusiasm for experiencing zero gravity.

Something far more alarming suddenly shattered the routine of Rayna's lesson. The once-mundane monitor displays, typically used for environmental data and communications, now show something far more alarming. One of the BioStat's information officers has overridden the usual feed, switching it to a live broadcast from Proxima, their distant home planet. The screen reveals an image of their asteroid, Jingai, but it's not the familiar sight they see on their security monitors. Instead, armed men in unfamiliar uniforms storming the facility, turning the place they call home into a war zone.

The students' excitement quickly shifts to confusion and fear. "What's happening?" a girl asks, her voice quivering with anxiety.

Another student responds, "I'm not sure, but it's kinda cool, right?" His unsteady tone betrays his uncertainty.

"Miss, is this a joke? One student demands, "Miss, is this a joke?" But Rayna, along with the rest of the class, remained frozen, unable to take her eyes off the screen.

Rayna then murmured," It's not me. I have nothing to do with this." Her voice was unsteady as she struggles to comprehend the terrifying events unfolding before them.

Proxima's media outlets broadcasted images of the distant asteroid, Jingai—a tiny dot against the vast expanse of space. Although Proxima is their home planet, most Proximians might have been unaware of Jingai's existence until now, when the safety surveillance system thrust the asteroid into the spotlight through its relayed footage. For those unfamiliar with Jingai, the images might appear confusing, but for those in the know, it's clear that someone intentionally shared this footage. Whether the motive is political, safety-driven, or journalistic, the goal is to ignite discussion and draw attention to the crisis unfolding below the isolated, pressurised can of air.

"Miss, what should we do?"

Like the others in the module, Rayna was unusually silent, her mouth slightly agape as she stared at the screen. The small, sporadic gasps from her rising chest were the only signs of her mounting anxiety. She knew she should speak, but the unfamiliarity of the situation left her struggling to gather her thoughts. The fear felt foreign and she found herself unable to control either her own reactions or those of her students.

The students were the first to act, slowly retreating past the orange-suits and making their way toward the exit. Once outside, they turned right towards the BioStat lift shaft, seeking refuge with their parents or, as teenagers do, in their bedrooms.

The mines' emergency alarms blared, but the monitors had already alerted everyone. Rayna, her gaze still fixed on the primary screen, noticed the secondary monitor's familiar scene had also changed. The comforting sight of the Centaurus, the company's ship, orbiting safely, was now marred by the looming presence of a second, menacing liner approaching from afar.

A shiver of unease crept down her neck as the faint hum of the air conditioning fanned against her skin. Her attention snapped back to the first monitor, where the live feed continued to cycle through various angles: from the mineral containers at the mine's transfer terminal to images of armed men in unfamiliar uniforms storming the facility.

The broadcasters were scrambling to cover the unfolding events.

"We interrupt our regular programming for an emergency update," the media officer announced, her voice tinged with visible unease. "There's a situation developing at one of our asteroids. Reports indicate that armed troops have entered the complex. We'll keep you informed as more information becomes available." Her attempt at reassurance was tempered by the uncertainty evident in her expression.

An assault on an asteroid, or any vessel in space, was a scenario far removed from anyone's expectations.

"Why?" Rayna questioned herself, her mind racing to make sense of the absurd situation.

"Why would soldiers want to attack us?" She tried to process the illogical scenario, speaking aloud to the last students who were exiting through the closing door. "They must be pirates... but why? I didn't think there were any pirates around here," she continued, as if the students were still listening.

The screen then shifted to a new image showing four more figures in the tubes, three men and one woman. They looked different from the others and for a moment, Rayna couldn't place why. Then she recognised the uniforms.

"Are they from Proxima? No, wait, they're from the Centaurus," Rayna realised.

The recognition offered some comfort, but it did little to reassure her amidst the chaos and confusion. The figure at the front of the group was massive, his size forcing him to constantly duck and weave through the narrow, curved spaces of the tubes.

To further complicate Rayna's already scrambled thoughts, an unfamiliar voice repeated over the mine's intercom system:

"Attention! Attention! Do not resist! No one will be harmed."

"Attention! Attention! Do not resist! No one will be harmed."

"Everyone must return immediately to their accommodation units."

In a sudden turn of events, the monitor pixelated and fragmented into a vibrant mosaic of coloured squares, before ultimately freezing on a screen that read "No Signal". The intercom continued to blare its commands, but Rayna was oblivious to them. The overwhelming situation and the flood of unexpected information were too much for her to process. Her mind, unable to cope with the rising tide of panic, rejected the blaring warnings and fixated instead on the blank monitor, now displaying nothing but the stark, unhelpful "No Signal" message.

Suddenly, a piercing THWACK! shattered the air as the tube's door slammed open, crashing against the suit rack with a deafening clang. The impact sent a ringing noise reverberating through the room. The door's hinges strained as it locked in place. Rayna's usual assertive posture collapsed. Her head instinctively shrank between her shoulders and her hands clenched involuntarily as the shock of the sound tore through her mind, sending chills across her skin like icy fingers. Stunned, she slowly turned toward the source of the noise, her expression blank with confusion.

"On the floor! Now!" a voice commanded, sharp and unforgiving.

"You, lady, on the floor, NOW!"

Rayna stood frozen, unable to move.

He grabbed Rayna's upper arm and spun her around, his gloved hand reaching out to seize her wrist. Rayna instinctively jerked her hand forward, but the soldier was quick, catching her thumb and twisting her hand forward. The sharp pain of the wrist lock shot through her, leaving her no choice but to comply as he guided her to the floor, firmly under his control.

Chapter 14

# THE GIANT

The tender docked onto the mine's airlock, its soft seal creaking as the hull adjusted to the mine's pressure. Kane's ears popped as the hatch opened and the team unloaded the supply crate, storing it in a nearby workshop among a stack of parts. The crate contained essential items, power packs, food, water, medical kits and radios. They never included weapons or operational gear in crates to prevent them from falling into the wrong hands.

Kane began the routine check-in. "Centaurus, this is Whiskey One, over."

There was no response.

"Centaurus, this is Whiskey One, over."

The pilot popped his head. "I'm getting nothing from you."

Kane adjusted his coms connections. "Centaurus, this is Whiskey One, over."

"Centaurus, this is Whiskey One, nothing heard, out."

The pilot attempted to relay Kane's message but received no response. While it was unusual, it wasn't alarming. There could be any number of technical issues at play. As a precaution, Kane gave the pilot new orders. "Seal the hatch. We'll signal when we're done. If were taking too long and the coms don't come back online, return to Centaurus."

The docking door sealed shut and the pilot begun reorganising the passenger hold before settling in.

It had been a while since Whiskey's last assignment to Jingai, so a lot could have changed since then. The purpose of this trip was to supply and recon, so the team headed down the tubes to reacquaint themselves with the mine's layout. No words were needed as the routine of reconnaissance came instinctively to the team and they knew exactly what to look for. Leading the way was Hulk, a figure of intimidation, a non-lethal deterrent that discouraged miners from delaying them. In a place where trivial entertainment was scarce, even making the security force wait could amuse the bored.

The mine buzzed with activity. Workshops reverberated with the clanking of machinery, mechanics laboured on hydraulics, technicians swapped out circuits and geological teams prepped their gear for fieldwork.

"This place is the asshole of the galaxy," Pope grumbled about the muck and filth in the confined spaces.

As they neared the atrium, where they would turn back and return to the tender, the inspection remained routine. A few mining vehicles were undergoing major overhauls, but nothing appeared out of the ordinary. They cleared access points of clutter and used the tubes as outlined in the mine's emergency manifest. They checked all emergency points and equipment for serviceability and made sure the fire doors' emergency passcodes were still enabled. This tedious assignment also allowed the team to become acquainted with the area.

Suddenly, the mine intercom crackled to life. "Attention! Attention! Do not resist! No one will be harmed. Attention! Attention! Do not resist..."

"What the fuck!" Pope exclaimed.

In an instant, the uneventful recon transformed into a possible hostile environment. Without a word, the team moved into an all-around defensive position. Drawing their pistols and scanning their arcs, some watching the front, others the rear. Training took over, leaving them poised and ready, waiting for Kane's orders.

Kane quickly assessed the situation. "An unknown voice... they sound serious... the liner Tyson is coming... they're likely from ... but why and where the fuck did they enter?" Questions flooded his mind. He evaluated them in milliseconds and made a decision.

"Alright, on me," Kane ordered. The team closed in around him, still facing outward with weapons drawn. They were all thinking the same thing.

"The visitors are probably from the Tyson."

"Sounded like he had an accent," Maihuni noted, picking up on the Earthen drawl.

"Yeah, I thought the same," Pope added.

Hulk asked, "What about the supply crate?Hulk was considering a return to the tender where the supplies were hidden.

"There's little in the crates that could boost our firepower. We're closer to the BioStat atrium. Better to hold up there if they're in the tubes, as a buffer for the BioStat community."

"Any questions?" Kane asked. There were none.

"Let's move before this turns into a shooting gallery." Kane knew they were under-armed for a prolonged fight.

Hulk led the way as the team fell in line behind him.

"Six-Three-Charlie, this is Whiskey One, over."

The pilot still had no communication with Whiskey One or Centaurus. However, he had heard the threatening announcements over the mine's intercom. With the tender sealed, he waited until he saw unfamiliar uniforms through the hatch portal. He didn't think twice in taking Kane's earlier advice and returned to the Centaurus.

Rayna had executed her role flawlessly thus far. The prospect of being caught in a situation involving weapons terrified her, but fear had a way of making her performance even more convincing. She wondered how proud her father would be. Now, her biggest challenge was maintaining this facade. Her last task was to gather crucial intelligence on how the miners would react to their temporary captivity, identify potential troublemakers and to uncover any plans to resist. At least she could take solace in knowing that all her students had escaped unharmed. Her hands had been loosely bound with cable ties by a soldier, who had discreetly advised her to maintain the illusion by keeping her hands within the ties.

Rayna had never been involved in criminal activity, nor had she experienced the riots that plagued the poorer districts of Proxima. She found herself in a situation she could hardly comprehend. Survival instincts urged her to comply with whatever she was told. She had no desire to resist. The soldier escorting her knew the value of her local knowledge and had orders to debrief her before returning her to the BioStat with the other captives. Rayna was acutely aware of the importance of maintaining this role, especially with the presence of many cameras that might inadvertently capture any slip-ups.

The soldier guided Rayna into a connecting tube that resembled a central corridor. She kept her head down, trying to avoid further chaos in her mind. The discomfort of having her hands bound behind her occupied her thoughts. Then, without warning, someone swept her feet out from under her. At first, she thought it was the door slamming shut behind her. The shock tightened every muscle in her body as she fell, instinctively rolling to protect her face. Her cap knocked askew, resting on the bridge of her nose. From the corner of her eye, she saw an imposing figure looming above her, his cheeks reminiscent of a Santa Claus without his beard. His height towered over an average man and his broad shoulders made his police gear look almost childlike in comparison.

Relief washed over Rayna as she realised the giant was focusing his unperturbed gaze on the soldier who had loosely bound her hands. The massive figure unleashed a sweeping left-handed strike, transforming her captor into a semi-lifeless puppet. The giant's hand tightened around the handle of the sheathed knife hanging from his waist. At the same time, his gloved hand clasped over the soldier's face, preventing him from collapsing onto the grated floor. With the soldier's face securely in his grip, the giant wedged the man's jaw against his skull, ensuring no cries would escape.

Without hesitation, the giant unsheathed his knife and drove it between the soldier's neck above his chest armour. Desperately, the wounded man reached out for his knife and swung with no clear aim, grazing the arm of another soldier. The giant twisted his blade, pulled it out and thrust it upward into his neck, right behind the jawline. The now lifeless body slumped, falling from the blade under its own weight. Rayna struggled to comprehend how the giant performed this brutal act with such ease. He gently lowered the lifeless body to the floor. Still gripping his face, he dragged it back inside the doorway.

Rayna was aware other soldiers were behind the giant, though she couldn't make out their faces, just an indistinct mass of armed men. Then a female soldier grabbed Rayna's arm, helping her to her feet and straightening her cap. The first gesture of kindness Rayna had encountered amongst these new soldiers. It was a small maternal act, offering comfort amid chaos. They shared no words, just hand signals. It was then that Rayna realised the constant noise of mining machinery had ceased, leaving an eerie silence. As the group moved towards the BioStat, unanswerable questions swirled through Rayna's mind.

"Why is this happening? Why are they killing each other? This was supposed to be simple, they said!" Rayna's thoughts raced, only to circle back to the cables binding her wrists. All she wanted was to flee.

The female soldier must have noticed the astonishment on Rayna's face, the look of someone on the verge of erupting with many questions. With a decisive gesture, her grip on Rayna's arm softened and she placed her finger over her own lips, signalling for Rayna to remain quiet. The signal was unmistakable and Rayna had no intention of tempting the consequences. These soldiers were resolute in their mission and not inclined to engage with any uninvited complications. They moved with methodical precision, securing their path through the modules until they reached the lift shaft leading to the BioStat.

One soldier spoke into his communication device. "Centaurus, this is Whiskey One, contact. Do you copy, over?"

Unknown to Rayna, the soldier's communications were still down. He was merely going through the motions in case Centaurus could hear him.

The tender pilot distanced himself from the mine, where he reestablished contact with Centaurus. The officer on deck immediately ceased all non-critical communications and ordered battle stations. "This is not a drill!" He then began trying to reestablish coms with both Whiskey and Papa, who were both now stuck on the asteroid and BioStat.

They entered the atrium, a room large enough to hold around thirty miners, shoulder to shoulder. On most days, only about ten workers waited there for the lift that would carry them up the central shaft to their quarters. The team immediately secured the heavy blast door behind them. The atrium was more than just a waiting area. It served as the structural base of the BioStat's shaft, doubling as an emergency shelter for miners. In their current situation, it provided a temporary refuge.

Whiskey knew this room well, having conducted countless search and rescue drills here. Its simplicity was its strength. It is always uncluttered, with no barriers, crates or furniture to trip over or hide behind. The walls were a dull grey, adorned only with maps of the mine's layout, safety notices, emergency equipment and a defunct communication terminal with an "out of order" sign slapped on it. The atrium had two entrances, the one they had just locked down, leading to the mine and a curved back wall that housed the lift. This layout gave them a significant tactical advantage. The fortified door they'd just secured was the only access point from the tubes, allowing them to concentrate their defenses on this single vulnerability.

Hulk guided Rayna to sit against the wall and she obeyed immediately. A group of her students stood by the elevator doors, anxiously awaiting its return. Those who had boarded earlier didn't wait for the others because they were too focused on themselves. The students left behind wore expressions of fear and panic as they were ushered next to their teacher on the floor. They complied without question, clinging to one another, their eyes wide with dread. Rayna's heart ached. She had never intended to harm anyone, least of all her students. Shame gnawed at her, knowing she had played a part in the turmoil. In stark contrast, the soldiers remained calm and unwavering, their attention locked on the mine and the lift, alert for any emerging threats.

Typically, human-made structures in space offered no glimpse of the other-side, but this reinforced door was an exception. At eye level, a small porthole provided a view into the mine's tubes, designed for emergency crews to see what awaited them on the other side. Pope positioned himself by the door, acting as a lookout while the rest of the team prepared themselves for what might come. Kane had his contact report ready in case coms were reestablished with the Centaurus, but when the red light above the lift's doors started flashing, accompanied by the repetitive 'Beerp!' sound, he held off. The lift was on its way down.

Hulk moved to position himself between the lift, Rayna and her students. The slight relief the soldiers might have felt earlier evaporated, replaced by a palpable tension as their fingers hovered over their triggers. Rayna's jaw trembled, confronted with the harsh reality that this might be her last moment. She glanced at her students, huddling together, trying to hide within each other's embrace. The lift came to a stop and the green light above the doors turned an ominous amber. The doors slid open, revealing a chilling sight, the muzzles of weapons pointed directly at them and behind those barrels, the wide-eyed faces of armed adversaries.

# Chapter 15

# THE ATRIUM

Sitting against the wall, peering over her knee, Rayna wished she were back in front of her father. She longed to hear his calm, confident voice, just as she had in her childhood. Rayna cherished his stories about their ancestors, who revelled in their ability to explore exotic places adorned with sky-piercing trees and vistas of unfrozen tidal waters. Her father's tales painted a picture of a family that had it all, with a history of triumphant business ventures and contracts secured through the artful greasing of the palms of corrupt politicians. Rayna held a deep affection for her father and believed wholeheartedly in his words. She could feel his unwavering loyalty to their earthbound family and his fervent determination to replicate their successes. Yet, she shook herself from this distraction, returning her focus to the reality at hand. Her eyes scanned the unfolding and unexpected events before her, absorbing all that transpired beyond her kneecaps. It was then that she realised her gaze had fixed itself upon the one who appeared to be in charge.

At the moment when fire and brimstone had seemed inevitable, the soldiers lowered their weapons, and the unwavering stances of everyone present softened into a collective sigh of relief. It became clear that they and the soldiers in the approaching lift knew each other. Rayna pondered how they had come to be inside the BioStat already, surmising that they must have arrived via a tender in the upper levels. The perpetual confusion did nothing to soothe her and she averted her gaze, turning her attention to the grey paintwork on the wall, yearning to vanish into its brush strokes if she could.

With the lift's doors now fully open, the first soldier exited the lift, entering the atrium, lowering his weapon and flashing a smile at the massive figure before him. He jokingly asked, "You cut yourself shaving again, big fella; you've got blood on your chin?"

Hulk's response to Papa One was predictably curt, "Fuck off!" as he nonchalantly wiped at his chin.

The ogre, who had loomed menacingly over Rayna, turned and drew his blood-soaked knife from its scabbard once more. He calmly lowered and positioned himself against the wall she stared at. His presence made Rayna shrink back, pulling her knees in as tightly as she could. She fixed her gaze on the tex-tured paint strokes on the wall, clinging to them as if they were the edge of a sheer cliff.

Behind her, she heard someone usher the students into the open lift. Rayna longed for them to wait, but it was clear they intended to put distance between themselves and the atrium. The lift doors closed, beginning its ascent, and she wasn't going with them.

"Why?" was the desperate thought that briefly consumed her.

The giant reached out and seized her trembling arm, carefully running the cold blade between her hands, separating the ca-bles with which her captors had bound her. Logic slowly over-came her panic and she realised that this formidable creature had just rescued her and her life was not about to end. She shuf-fled sideways, creating a bit more space, then nervously tucked away a stray strand of hair that obstructed her view. At that mo-ment, her appearance was the last thing on her mind.

"Not a professional job with those cable ties," he remarked, sharing with the others his observation of the captors' lack of professionalism.

Recent events had clouded Rayna's understanding of what was unfolding. Was this man beginning to see through her innocent facade, growing suspicious of the deeds she had truly committed? Her fear returned in full force when he locked eyes with her, remarking, "You could have easily slipped your hand out of those if you'd tried."

Had he uncovered that she wasn't as naive as she appeared? In response, she swiftly shifted the conversation to divert the giant soldier's thoughts away from the cable ties. A sly smile played on her lips, carrying a subtle air of justifiable pretentiousness. It wasn't an image that was easy to project under the current circumstances, but it had always proven effective in swaying thoughts towards her beauty. The giant was indeed male and the tactic worked. And so, for the time being, she remained in the atrium. Was it just so he could release her from the ties? She didn't know why, but it appeared she now belonged to those who believed they had rescued her.

His massive hands nearly enveloped his beret as he lifted it from his head, unveiling a thick head of tightly coiled curls. The release of pressure caused him to emit a tremendous sigh of relief. "I still can't believe they can make caps big enough to fit that noggin." Came from someone.

Rayna found herself equally captivated by this man's carnival appearance. Her eyes locked on his every movement. He remained seated beside her, but he now seemed to pay her little attention. A nondescript but thick moustache and a prominent mole just above his left eye accompanied his mop of unruly curls.

"No one would have dared to make fun of that mole when he was growing up," she thought. He sheathed his knife, assessed his surroundings and relaxed. Within a moment's thought, he started searching his pockets and gave up when he couldn't locate what he was looking for.

"Feeling hungry?" he asked Rayna casually, having mastered the art of staying calm during long stretches of downtime.

Rayna shot him a puzzled glance. How could he possibly be thinking about food right now? She wondered. Hulk caught her expression and mistook it for hunger. His face was an open book, clearly focused on one thing—food. She shook her head in response.

In the centre of the atrium, a group of soldiers chatted quietly while others sat around the walls, their eyes periodically watching the soldier looking through the door.

"I'm starving." The giant declared, oblivious to her confusion. "I could really go for an entire pot of braised spinach and chickpea stew right now."

Pope glanced down through the door's observation portal when Maihuni tapped his leg, raising an eyebrow as she subtly gestured toward Hulk and Rayna against the wall. "Looks like Hulks got that lost puppy syndrome again," she said, just loud enough for Rayna to catch the warning in her voice about Hulk's overbearing nature.

"Hulk... that's his name." Rayna thought, as he pulled out a wrapped sandwich from a pocket.

"Don't mind the thick-necked oaf," Maihuni joked to Rayna, her tone friendly despite the teasing. "He's got to keep himself well-fed to balance that enormous head of his."

Hulk took the bait, his attention snapping to the banter. He was up for some good-natured ribbing to pass the time.

"You wouldn't have trouble wrapping those legs of yours around these hips!" Hulk shot back, clearly proud of his bulk.

"Fuck off!" Maihuni replied with a smile, acknowledging her lack of a snappy comeback.

Kane was in the centre of the room discussing the situation with the team leader from SERT Papa. Turning and seeing Hulk, he couldn't resist a comment. "Fuck me Hulk! Is there any time that you don't think about shovin' food in your face?"

"Nope," Hulk answered with a mischievous grin, completely unbothered by the surrounding events.

"What's your take then?" Kane knew that Hulk often had a different perspective, one that could be valuable. After all, he was the one who had taken down the soldier escorting the woman and his observations may provide crucial insight. The two had been in the same squad for years, building a bond that went beyond the usual camaraderie. It was a friendship defined by mutual respect and professionalism. They both understood that getting too close could cloud their judgment and bring heartache if things went wrong. So, the friendship remains at joking taunts and drunkard binges.

"The grunt had the same gear as us, but he was armed to the teeth." I grabbed this off him." Hulk held up the SALUR—Small Arms Laser Urban Rifle, knowing it was useless without an authorised handprint. Still, it was one less weapon for the enemy and Hulk intended to render it completely unserviceable.

"I noticed he had removed his unit badge, but there were some light blue threads still stuck in the velcro," he added.

"Earth?" Kane guessed, knowing Earth unit badges typically had a light blue background.

"Probably." Hulk replied. "Oh, and his gear was coated in black dust," he mentioned with a shrug, as if it were just another odd detail.

Kane turned back to the SERT Papa leader, whose task was to secure the BioStat. Papa One had observed the situation on the monitors up in the BioStat, watching the invaders and Kane's team make their way to the atrium before the screen went offline. He expected that there might still be miners attempting to enter the lift and it was logical to assume that Kane's team would be in the atrium.

They assessed the situation carefully. The enemy's numbers, tactics and motives remained unknown. They had probably originated from the Earthen liner, the Tyson. The only certainty was that Centaurus' mission and by extension theirs, was to protect the mine and its workers.

"It's likely they're from Earth, especially with the Tyson so close and those blue threads Hulk found." Kane began.

"Agreed. But how did they get to the mine? The Tyson isn't anywhere near the asteroid and we have detected no shuttles from her." Papa One pointed out.

Thinking it through, Kane said," That's the big question. But it doesn't change our objective."

"It might, if they get into the BioStat the same way they got into the mine." Papa One countered.

"Good point." Kane acknowledged, realising that holding their ground might be a losing strategy. If the enemy could infiltrate the BioStat, they would cut off the atrium entirely.

"They'll most likely be planning to blow this door. So what's your call, boss?" Papa One asked, aware that the final decision rested with Kane.

The enemy had the upper hand. Not just in firepower and likely numbers, but more critically, in intelligence. "They know what's happening and we don't. Until we have a clearer picture, I say we fall back and secure the BioStat."

"Agreed," Papa One replied, pressing the button to call the lift.

Kane took the moment to reach out to the Centaurus."Centaurus, this is Whiskey One. Contact report, over." No response.

"Hey, boss!" Pope called from the door.

"We've got company. Six. Maybe more. Heavily armed." Pope reported, his voice steady despite the tension. He gave the enemy a cheeky wave, a bold gesture meant to taunt. Professionals like him brought colour, flair and confidence to the battlefield. Without it, combat would be nothing more than a brutal clash of soldiers.

Pope watched as an enemy soldier casually walked up to the door, leaned over and squinted into the portal, trying to see who else was inside. Pope, standing like a human barricade, obstructed his view and flashed a grin that was purely comedic.

Pope hit the intercom button. "Bet you didn't expect to run into us down here."

The enemy, without skipping a beat, shot back, "I'm glad you're here. You're outnumbered, so open the door."

Pope pretended to consider it. "Well, if you and your pals are planning on surrendering, we're not yet ready to accommodate you right now."

The enemy ignored the humour. "Your outgunned. Even if you don't open it, we're coming through that door."

Pope shrugged, still grinning. "I've offered you a simple solution. Might want to take a moment and rethink your options."

With that, the enemy turned and withdrew.

"My guess is these guys are an advance team," Kane said, his tone measured. "The rest may wait for the Tyson to arrive before making their move."

Kane quickly gathered both teams at the centre of the atrium, addressing the situation with calm authority. "Listen up. We don't know what the enemy's endgame is, or how they infiltrated the mine. They're fully armed and prepared for a fight, which we can't match. There's also a chance they're bypassing us and heading into the BioStat above. The people up there are our priority." He pointed towards the lift. "We're moving up to the BioStats. Whiskey team will defend the nearest BioStat and Papa will secure the outer one. Specific details will come once we've assessed the new surroundings."

Kane then emphasised the mission, making sure everyone was clear about their objective. "Our priority is defending the BioStats and protecting its people. We'll hold until we receive further orders."

He continued, "Whiskey and the female will move out as soon as the lift arrives. Papa, you'll hold this position until the lift returns. If there are any more miners in the tubes, they won't be coming through that door. That door is to remain sealed." Kane's experience and composure in chaotic situations shone through.

"Any questions?" He offered.

The silence was only broken by the sound of the approaching lift and everyone preparing to move out.

Chapter 16

# A SMALL TATTOO

Waiting for the lift felt like being a cornered animal waiting for the inevitable predator to pounce, with everyone anxiously eyeing the indicator light, hoping to see it switch to amber. Kane glanced around the atrium, catching sight of Hulk, who was deep in conversation with the woman he was escorting. Hulk, ever the disciplined soldier, avoided the instinct to point, a habit drilled into him to keep a low profile and avoid drawing attention to the leader. Instead, he subtly nodded in Kane's direction.

"Don't ask me, lady," Hulk deflected the woman's question, passing the responsibility up the chain. "Ask the boss over there."

Kane wondered what trivial concern she might have, considering now wasn't the time for unessential conversations. In his experience, police personnel were trained to think as a unit while civilians often prioritised self-interest. Still, he maintained a respectful demeanour about the chances for a fruitful conversation.

When Kane's eyes met hers, a wave of attraction surged through him, threatening to eclipse his better judgment. Staying focused, battle-ready, became a struggle as primal urges blurred the edges of his resolve. She was the kind of beauty that could easily turn into an obsession, a pursuit, even a perilous distraction. Kane had encountered situations like this before, but they being rare and always risky. Danger was a constant companion in his life, almost justifying the pull toward such an enticing distraction.

Then, reality struck. As the adrenaline rush faded, he became aware of the pain from a wound he'd previously ignored. The knife wielded by the soldier Hulk had taken down had scraped off Kane's vest and pierced him beneath his shoulder plate. Waiting for the lift provided a brief pause. His senses wound down, allowing his body to remind him of the injury. He gingerly touched the area, feeling a sharp sting as his fingers brushed against a small trickle of blood. Grimacing, he made his way over to Rayna and Hulk.

"Move aside, Hulk," he ordered. The stark size difference between the giant and his leader still captivated Rayna.

"Pussy," the big fella said with a smirk, referring to Kane's obvious minor injury.

Rayna's gaze shifted away. Recognising this wasn't the moment for deception.

Maihuni stepped forward when she saw Kane's bloodied fingers, prepared to offer help. Her expression stayed calm, guided by a quiet sense of duty, like a mother caring for a child's scraped knee. Though every team member could manage minor injuries, Maihuni had a particular talent. She was always the first to act and the most skilled at providing care. She had once been a nurse and excelled in her field. But for reasons only she understood, it wasn't enough. The hospital routines and predictable shifts began to feel restrictive, like walls closing in. Beneath her calm exterior, a restless energy simmered, pushing her toward something beyond the boundaries of medicine. A call to serve in a way that defied the life she had so carefully constructed.

As the team's maternal figure, Maihuni carried herself with a quiet strength that made her presence both comforting and formidable. She was a slim woman in her thirties, her naturally black hair cropped short on the back and sides, with a long fringe that framed her face. Her well-toned abs and golden skin were a testament to her rigorous dedication to fitness, a discipline that mirrored the precision with which she approached every task.

Her demeanour was predominantly serious, a stark contrast to the lighthearted banter that often filled the team's downtime. While the others laughed and joked, she remained focused, her dark eyes always scanning, always assessing. Maihuni's smile was rare, a subtle quirk of her lips rather than a full expression of joy, more often replaced by an exasperated roll of her eyes when the others' antics grew too boisterous.

When she spoke, her words were few but carefully chosen, sharp enough to cut through the noise and get straight to the point. There was no room for unnecessary chatter in her world—only efficiency and purpose, qualities that made her indispensable to the team. Even in the most chaotic situations, Maihuni exuded a calm authority, a steadying influence that the others relied on, whether or not they admitted it.

Kane winced as Maihuni administered a hemstat injection.

"Did you see what uniform he was wearing?" Kane asked Maihuni, seeking her insight.

"Yeah, but I don't get why anyone would wear it unless they were from Earth," she replied, applying staples to his wound.

Rayna's heart raced. She thought her connection to the Earthen soldier was dangerously close to being discovered.

Kane pondered aloud, "The question is, how did they get here before their liner?" It was a piece of the puzzle that none of them could figure out.

"When I dealt with that guy..," Hulk began, Rayna flinched, still haunted by the memory, "...my gloves ended up being covered in black soot. That stuff only comes from the mines."

Kane mulled this over. "Even if they were in the mines, it still doesn't explain how they got there."

"Maybe they've got a shuttle on the far side of the asteroid," Hulk suggested, his attention still on Maihuni's work.

"Not likely," Maihuni dismissed the idea. "If they used a shuttle, the Centaurus would've detected it long before we lost coms."

Rayna discreetly lowered her gaze, still trying to disappear.

"Doesn't matter too much. There here." Pope added.

While treating the wound, Maihuni noticed Rayna was tidying up her appearance. As she pushed up her sleeves, a small tattoo became visible.

"That tattoo looks different," Maihuni noted, more curious than merely observational. She had seen something similar before but couldn't place it.

Kane also glimpsed the ring tattooed under Rayna's forearm, but thought little of it.

Rayna, however, seemed keen to avoid further attention on her body art, offering a quick smile as she subtly concealed the tattoo.

Unfazed by Rayna's reluctance to talk, Maihuni pressed on. "In SERT, we all have tattoos," she pointed to the area above her inner right elbow. "It's tradition. We get our blood type marked here, just in case."

Rayna changed the subject. "Are you okay?" she asked Kane, trying to divert the conversation.

Kane smiled bravely and winked, determined to maintain his composure in front of this beautiful woman, even if the wound had been more serious than it was.

Maihuni wasn't so easily deterred. "What's your tattoo about?" she asked again, her curiosity piqued.

Rayna spun a story about a wild night at the end of engineering school, where everyone got their course logo tattooed. It was believable enough and Maihuni's interest quickly waned.

After finishing the wound dressing, Maihuni stepped back and nodded with a touch of satisfaction. "All set, boss," she said.

Kane gave her a nod of thanks, gritting his teeth against the lingering pain as he began suiting up in his body armour. He secured his headset, the small click of the fastening signalling his readiness.

The lift doors slid open with a soft hiss, Kane straightened and his focus sharpened. Hulk hoisted himself up and extended a hand to Rayna, effortlessly pulling her to her feet and helping her into the lift. As she stepped inside, the confined space trapped a suffocating wave of stale, pungent sweat, the kind that clung to the heavy armour and gear they all wore. The smell hit her hard, nearly making her gag. She understood it was inevitable, considering the gruelling conditions they endured, but it didn't make the stench any easier to bear. She couldn't help but wonder how Maihuni tolerated it day in and day out. Rayna clasped her lips, trying to breathe as shallowly as possible, but the odour seemed to seep into her very pores, growing stronger with each passing second. The lift's walls seemed to close in around her, amplifying the assault on her senses and she silently willed for the last person to enter to end this ride.

Kane, the last to enter, noticed her discomfort among the soldiers. She seemed out of place among those taller. Tall stature being an evolutional trait of many Proximians. Rayna squeezed her nose in a futile attempt at surviving the trip to the BioStat.

"Cute," Kane thought, watching her discomfort.

As the lift doors closed, the muffled beeping gradually faded. Pope broke the silence. "Why is it that people always stop talking in a lift?"

Maihuni shot back a subtle counter, "You just contradicted that," wondering why he felt the need to fill the silence.

Hulk, as usual, changed the subject entirely. "It would be nice to have some music right now?"

The grinding and humming of the lift shaft replaced the awkwardness and then, with a loud clank, the mechanisms locked into place, spinning them upwards 'Beerp!' The buzzer signalled their arrival as the lift approached the outer doors of the first BioStat.

As the doors opened, Rayna instinctively moved closer to Hulk, sensing him as her best chance for protection. Despite the off-putting lift ride, she remained with the group of soldiers, not just for safety, but to pursue her unfinished agenda.

# Chapter 17

# WELCOME

From the edge of the lift's door, an elderly man peered out carefully. "Welcome, welcome," he greeted, though his trembling voice betrayed his anxiety about how the lift's occupants might respond. "I'm Dakota, the BioStat's administrator,"

"Don't worry, Dakota. We're from the Centaurus," Kane reassured him, prompting a visible sigh of relief.

"What's happening? We've had miners and minors, sorry, miners and students, coming up the lift with many wild stories," Dakota remarked. His concern was clear.

Kane responded with a calm but urgent tone, "Who knows? The real question is, what are you and I going to do about it?"

The team quickly assessed their surroundings. "We'll set up our office right here by the lift. We have another team in the can above us. I have a few tasks for you, Dakota," Kane instructed. "First, go find your electrical engineer and systems officer, then come straight back. We need to act fast to keep your people safe."

"Systems officer?" Dakota asked, confused.

"Whoever handles the cameras," Kane clarified, pointing at the one above his head.

"Got it," Dakota replied, turning to leave.

"Wait," Kane stopped him with a hand on his shoulder. "There's more."

"Yes?" Dakota turned back, attentive.

"Ask around if anyone has experience handling weapons and have them report to me. Also, bring some food and water," Kane added.

Dakota hesitated for a moment, then nodded and set off, briefly considering which direction to take before scurrying away.

"Poor old fella might have a heart attack with all this excitement," Hulk shared with Kane.

Kane positioned himself in front of the lift, keeping an eye on the signal lights above the doors. He knew they had some time before the enemy breached the atrium.

"Hulk, scout for alternative exit routes in case we need to evacuate," Kane ordered.

"Huni, find the Medical Bay," he added.

"On it," she responded, already preparing for the possibility of casualties.

The signal light above the lift showed the letter M as it arrived below to transfer Papa to the second BioStat. Miners poured in, eager to offer their help. Among them was Davo, who spotted Rayna in the corridor. Though he was relieved to see her away from the mine, her withdrawn demeanour was unsettling. This wasn't the Rayna he knew. He also knew this was not the time to chat with his drinking partner and focused on supporting the soldiers.

At twenty-eight, Davo stood tall at over a hundred and ninety centimetres, his athletic build a testament to his active choices in life. His dark brown hair, mostly hidden under his cap, swept over his left eye. With a strong, angular jawline, he exuded intensity. Confidently wearing a durable, form-fitting jumpsuit, his utility belt bristled with tools and gadgets essential to his trade. Davo's personality mirrored the energy of the circuits he worked on, making him a natural leader in high-pressure repair situations. As the station's electrical engineer, he took pride in keeping everything running smoothly.

Dakota returned, spotting Davo and quickly approached him. "Ah, there you are. Come with me. We've got a job for you." Davo glanced at Rayna once again and nodded, understanding that he was required elsewhere.

"This is our electrical engineer," Dakota introduced Davo to Kane. "He's the best around. Seen none better."

Kane shook Davo's hand and asked about the systems officer.

"You've just met him." Dakota pointed at Davo, who was also a skilled multitasker.

"I need you to disable the lift temporarily and any tenders attached to this BioStat." Be ready to reactivate them if needed, then report back," Kane instructed.

As Davo moved to carry out his task, Kane noticed him exchange a glance with Rayna down the corridor. Kane's gaze lingered until Rayna caught him looking.

"Oops!" Kane thought, refocusing.

"Dakota, I want the portholes manned to keep an eye out for any approaching tenders. The Centaurus can't notify us and I'm sure they missed how the guys downstairs got here," Kane continued.

"Understood," Dakota listened closely, absorbing the instructions.

"Round up all the unassigned individuals and notify them about a security issue in the mine." That's all we know." The invaders are in Earthen suits, blue and grey. Ensure that everyone knows our uniforms are brown and grey. Anyone not helping should stay in their modules. That's their safest place." He added, "Organise galley trips if needed, but tell them to be prepared to stay in their modules for a while, so try to relax."

"Got it," Dakota waited for further instructions. He had learned from their last conversation.

"That's all for now. Thank you, Dakota," Kane concluded, clarifying it was okay to leave.

Two BioStat personnel arrived, both with prior experience as cadet officers during their teen years. A program designed to recruit future police officers.

"Ever fired a SIPLAT? (Standard Issue Pistol - Laser and Taser)," Kane asked. They both acknowledged. Kane added, "We don't have extra weapons, but be ready if one of my team members needs your support." He configured his SIPLAT to recognise their handprints and stationed them as lookouts at the corridor ends, instructing them to stay close.

Kane called out. "Pope, when the team returns, make sure these two have weapons clearance."

Hulk returned with Davo. "I found Sparky here messing with the tender bay controls. He knows his stuff," Hulk reported. Davo now had his tablet in hand.

"Sparky," Kane addressed him by his new nickname, already forgetting his real one, "Can you manually shut down the Bio-Stat's lift and cameras while keeping the mine's cameras on?"

Davo replied, "Whoever's in the mine, they've already tampered with our monitoring systems. It's easy enough to access the cameras. I can isolate the BioStat's, but the mine cameras went offline a while ago. I can no longer control them."

"Alright, do what you can. Make them as blind as we are and get that lift offline," Kane instructed.

Davo sat down by a wall next to Rayna and began working on his tablet. It wasn't long before he reported back to Kane, "These guys are good... they've blocked all access to the lift's software. I can't crack it."

# Chapter 18

# KILL ZONE

Kane and Pope shared a brief, knowing glance before turning their focus to the task ahead. Kane quickly analysed the options before him, weighing the risks and benefits of each. Once he had a logical strategy, he began discussing his plan with Pope. They quickly ruled out using the lift itself as a kill zone. Its limited firing range and vulnerability to explosives made it too risky. Instead, they pinpointed the lift's threshold as the optimal target area, offering better coverage and control.

"This is where we'll set the kill zone," Kane outlined, mapping out his strategy. "We'll split into two groups further back along each of the two corridors, using whatever cover we can find." The corridors stretch across the diameter of the BioStat, with the lift at its centre. They completely encircled the lift shaft before continuing on the other side, providing the primary access point with ample lighting and minimal obstructions.

"This setup keeps us at a safe distance with clear retreat routes," Kane explained, before turning to Pope. "Your thoughts?"

"How are you going to lure them out from the lift into the kill zone?" Pope asked.

"We don't need them all to enter the corridor," Kane clarified, pointing to the fire lanes. "Our goal is to prevent their access. It won't be a high body count, but it will stop them from advancing further. Also, we don't know what kind of assault gear they might bring up in that lift."

"Got it. This could turn into a fighting withdrawal," Pope realised.

Without delay, they began executing the plan. "Alright, everyone, gather around," Kane ordered, rallying the BioStat helpers.

"We're throwing a surprise party and you're the welcoming committee." His commanding presence holding everyone's attention as he delegated individual tasks.

"Dakota," Kane said. His old self returned while taking charge, doing what he knows best.

"How can I assist?" Dakota asked, aware of the impact this situation could have on his community.

"Get some of your people to gather large crates and position them on both sides of the lift door." Kane instructed, pointing to the spots just out of sight from the lift.

He led Pope to the spot where he and Maihuni would set up their barricade. "Make sure Maihuni's ready when she gets back," he instructed.

Turning toward the lift, Kane raised his arm, marking out the firing lane. "About eight meters down the corridor, just a meter ahead of where the lift opens into the well-lit area."

He paused, thinking ahead. "If there's any extra crates, stagger them along the corridor—just in case we need to fall back."

Hulk received similar instructions for setting up crates in the opposite corridor, where he and Kane would be stationed. The former cadets were split between the two teams, authorised to use their SIPLATs of each team member in case someone falls.

"Thanks all, now let's do this!"

"Shields would be nice right about now," Kane thought. Carrying shields hadn't been practical or even considered. With only their personal weapons, there was no use dwelling on the impossible.

With the plan set, they started organising the crates as they arrived. As sandwiches and water arrived, Rayna kept herself occupied by distributing them to those waiting for the next batch of crates. Hulk wasted no time, quickly grabbing a bite to eat, and soon after, Kane and Pope joined him. Maihuni returned from the Medical Bay just as Dakota shuffled in, ready for his next orders. With everyone gathered, Kane took the moment to reinforce the plan.

"Two fire teams, with backups ready to step in if needed," Kane outlined. "If the lift moves up from the mine, everyone stays silent and hidden behind the crates. Once they step into the kill zone, I'll fire the first shot." He made that clear. "Once I fire, everyone else should get up and empty their weapons into the target area. That's when the chaos begins, so no one move a muscle before then!"

"If anyone goes down, the cadets are to take over their weapons and keep firing," Kane continued. "Dakota, Rayna, I want you two to tend to the wounded or handle any other needs." Rayna nodded. The confusion in her mind would agree to almost anything.

"When I raise my weapon, cease fire," Kane instructed, demonstrating by pointing his weapon skyward. "If I'm hit, any of you can raise your weapon." He continued, "After the ambush, I'll give one of two commands, 'hold' to maintain positions or 'break' to evacuate to the tenders."

Rehearsing a withdrawal wasn't possible, so Kane turned to Hulk, asking him to outline the route to the tenders and how they would pick up the BioStat personnel along the way.

Finally, he addressed Dakota. "As the lift rises from the mine, inform your team leaders to notify everyone to be ready to evacuate their quarters. No luggage."

"Will do," Dakota responded.

"Sparky, once the lift moves, make sure the mines tenders are ready to go," Kane directed.

"Got it," Davo replied with a nod.

"Remember, SERT Papa and the remaining personnel are in the BioStat above. I'm the only one who will trigger the ambush. However, if you need to engage, do so only when the targets are clearly hostile," Kane reiterated.

He added one last command. "If the plan falls apart and you can't hold your position, gather everyone and evacuate."

"Questions?" Kane asked, his voice steady. Silence followed.

As expected, Kane then launched into a motivational speech to wrap things up. Hulk and Pope shared a knowing smirk.

Pope leaned in and whispered, "Right on cue."

Kane ignored them. "None of us are here to die for our planet. Those bastards are here to die for theirs. So, if shit hits the fan, bug out."

Chapter 19

# FADED FRECKLES

Everyone dispersed to their tasks. Once the food service finished and the BioStat's kitchen staff collected the trays, Rayna took a moment to rest. Unexpectedly, Kane approached her, his manner suggesting he had something important to discuss. Or perhaps he was responding to the occasional curious glance she had cast his way. Being caution with his injury, he carefully lowered himself to sit next to her on the floor. Rayna turned to face him, meeting his gaze, which was a soothing shade of blue. Far less intimidating than the sharp, unwavering stare of his towering companion.

"What did you say your name was again?" Rayna asked.

"I didn't," Kane replied, still trying to wind down enough to take a break.

"Sorry, I guess you didn't," Rayna admitted. Then involuntarily, she blurted out, "Who are you people?" Realising her abruptness, she quickly added, "Oh and my name's Rayna."

Kane extended his hand with a friendly smile. "I'm Kane. Pleased to meet you. We were checking the mine when our paths crossed." He paused, waiting for her to respond, but she just stared at him. Undeterred, he continued, "Sorry, it couldn't be under better circumstances."

It then dawned on her that their paths had crossed before, but under circumstances she'd rather forget.

**109**

Kane noticed Rayna looking confused and said, "You'll be fine." His soothing tone momentarily reassured her as it had before.

"Oh, am I staring?" Rayna wondered, feeling the urge to keep eye contact. But then she looked away, her thoughts drifting to her family back on Proxima. She smiled faintly, realising that Kane had the same comforting presence as her father. Yet, she couldn't ignore the attraction she felt toward Kane, despite knowing it was dangerous territory.

Exchanging names felt like a significant step for Rayna, almost like she was crossing a line. She knew she needed to gather her composure to extract potentially valuable information, especially if the invaders took control of the BioStat. Yet, this newfound connection made her question her motives for continuing her role as a spy. The danger was something she had never experienced before and this group of warriors were unlike anything she had imagined. Their camaraderie, their deep bond, it was all so new and she felt an unexpected pull towards it.

Questions raced through her mind. Was she driven by a need to survive or by a desire to prove herself? Had her father's influence really shaped her path, or was she simply following the course he had set? Was her dream of reaching Earth truly her own, or was it just an echo of his ambition? Despite these doubts, her respect for her father was a bond she could never sever and she clung to the hope that her actions would make him proud.

Rayna watched Kane closely as he adjusted his beret, revealing his short hair. She glimpsed the nape of his sweaty neck, his hair fair, possibly ginger. He seemed of average height with a lean, fit build, not thin to the point of being unhealthy, not overly muscular, but clearly strong. Kane's exposed forearms and neck gleamed with sweat, his toned muscles a clear testament to his dedication to physical fitness. More so than most men his age, Rayna thought. She couldn't help but admit to herself, I appreciate men who are physically strong.

She shyly glanced at him as he removed his armoured vest, noticing his light complexion marked by scars and blemishes. His skin wasn't as smooth as most Proximians, but it seemed to bear the distinct marks of a life shaped by the rigours of his chosen profession. Faded freckles dotted his arms, making him appear more approachable, more human. These freckles, once the target of childhood teasing, had now softened with time, adding to his rugged yet attractive appeal.

Her eyes tracked his as he scanned the room, his expression a blend of weariness, focus, with a slight distraction from the knife wound. A conflict stirred within her as she pondered the possibility that he possessed something that piqued her curiosity, yet it remained elusive. She felt a strange attraction to him, one she couldn't quite understand. His appearance was unremarkable, yet there was an unfamiliar aura about him that intrigued her.

"Your friends really respect you," she remarked, her fingers brushing her lower lip as she spoke.

"Thanks. Partly because I'm their boss, but that respect is more from being a team."

The concept of a team was foreign in Rayna's civilian experiences, prompting her to ask more about it.

Kane explained, "It's like a family. Unless you're part of our family, it's hard to understand how much we rely on each other and the lengths we'll go to protect one another. It's forged through our tough training, loyalty and a shared sense of honour." He smiled and added, "...and often solidified with alcohol."

"Got it!" Davo suddenly exclaimed, breaking their conversation. He raised his hands in triumph, shaking his portable workstation like a prized toy.

"Got what, Sparky?" Kane asked, glancing over.

"The monitors are still down, but I've isolated the blockage in our transmission signals. You're all set to proceed!" Davo replied.

Kane reached for his radio set, which had fallen to the floor nearby." In a hurry, he clumsily placed it on his head and tapped the coms button. "Centaurus, this is Whiskey One, over."

"This is Centaurus. Send over," came the response.

"This is Whiskey One, contact report, over."

"Send, over."

"Centaurus, Whiskey One. We've made contact with Papa One... Hostiles in the mine... Numbers unclear... Hostile entry method unknown... No friendly casualties." Kane relayed, careful not to reveal their exact location.

"Copy that, over," Centaurus responded, awaiting further details.

"Hostiles are well-equipped... We've regrouped at Papa's designated insertion point... out," Kane provided more information, indicating their shift to the BioStat.

With the XO now updated, Kane shifted back into mission mode. He made better efforts at securing his headset and returned to his position behind the crates.

"Great job, Sparky," Kane praised Davo for a job well done.

The Centaurus' command centre worked to gather data and evaluate responses. Kane knew his team's role was crucial, understanding the broader implications of their actions. The Tyson remained distant, steadily advancing but unresponsive to communication. The mission for the Centaurus was clear: protect the mining community and the mine itself. They would adjust their strategies as needed and if the XO had more intel, Kane would welcome it.

He walked to the other barricade to update Pope and found entertainment in watching him authorise his SPILAT with one cadet.

"Pass over your wanking spanner," Pope said nonchalantly, his words slipping out with the ease of someone used to speaking with his comrades. The former cadet shot him a disgusted look.

"Your firing hand, your trigger finger. Unless you're ambidextrous? In that case, you're lucky," he clarified. Pope unintentional using language more suited to his comrades.

"Oh, I get it." She giggled, placing her hand on the pistol grip to register her prints.

Chapter 20

# THE JINGAI VOLUNTEERS

The BioStat residents worked quickly to deliver crates, using slimline forklifts for the heavier ones. With protection in mind, the team arranged the crates carefully, stacking them to create a clear line of sight toward the kill zone. The team staggered additional crates down the corridor, forming a zigzag of obstacles. They detached wall panels from various locations and repurposed them to strengthen their primary defenses.

Pope directed lookouts to monitor portholes, setting up early detection measures for enemy activities. While the Centaurus provided overhead protection, restrictions on spacecraft weaponry meant it could not engage in combat, relegating it to the role of an observer. SERT Papa was likely taking similar steps to secure the BioStat above, ensuring that all entry points were monitored and all defensive measures were in place at the lift. The enemy using that route appeared unlikely, especially with the Centaurus monitoring the area and able to spot any hostile shuttles approaching. Despite the low probability, every precaution was being taken, leaving no room for error.

Dakota was taking a break in a recessed doorway at the far end of the corridor when Kane approached. "Dakota, I need you to grab two sturdy people. I've got a job for them."

"Sure," Dakota replied, doing up his jacket and shuffling away.

The resounding clatter of hydraulic lifters and the squeak of rubber wheels continued as the final crates were being manoeuvred into place. Kane and Pope coordinated the movements, while the rest of the team took up positions behind their barricades. Space was tight, leading to occasional collisions and a few exchanged curses.

"I can only assume the enemy is from the Earthen ship. There's no other explanation," Kane speculated.

"Yep, all because of that bloody lottery ticket," Hulk chimed in.

"Really? You think that's the reason?" Kane asked, considering the idea.

"They could send more troops our way. Who knows how many?" Hulk added grimly.

"The higher-ups would be aware and making contingency plans," Kane noted as he directed the next crate delivery.

Hulk recognised Kane was vocalising his thoughts. "Yep," he simply replied, waiting for further developments.

"We can't base our defence on unknowns. Is there any way they could enter besides the lift?" Kane wondered. What he knew for certain was that the enemy had them outgunned.

"If they can sneak onto this rock that easily, they could have an entire army down there for all we know," Hulk pointed out.

"Yeah, but something is missing," Kane murmured, sensing he had overlooked something crucial.

"There sure is. How are they gonna feed all those troops?" Hulk questioned, prompting Kane to shake his head at Hulk's food fixation.

"I swear you say that shit just for a reaction," Kane paused, then realising he might be onto something. "... or maybe not."

Hulk continued, "Think about it. The enemy needs logistics—food, water, ammo and medical supplies. We are stopping them from getting that shit from the BioStat, so they're only going to get that from a liner. Which means there will be a lot more activity from the Tyson."

Kane remained vigilant, his eyes fixed on the lift lights, anticipating any change. Shifting his weight to ease the slight pain, he felt a light breeze brush against his arm, causing his hair to stand on end. His first thought was an air leak, maybe a pipe punctured by the movement of crates. But then he noticed the source of the breath of wind was coming from Hulk.

"What the hell do you have there?" Kane inquired, puzzled.

Unzipping his jacket further, Hulk replied, "I found him near the farms. He looked like he was hungry, so I thought he could use a snack." Hulk pulled out a lettuce from his thigh pocket and removed a leaf, which the white rabbit eagerly began munching on. "So instead of feeding it at the farm, you took a frickin' lettuce, too?" Kane remarked.

Hulk's hand slipped into his jacket, retrieving the floppy-eared rabbit. Though it seemed tiny within his oversized hand, it was of normal size. Typically, an animal this small might display signs of shock when handled by such a large person, but it seemed unfazed, content with its new companion.

"I thought it would make a good mascot," Hulk explained with a smirk.

"What, the lettuce?" Kane quipped.

"The rabbit, smart arse."

"Hulk, where the hell are you going to keep it?" Kane asked, his disbelief evident in his voice.

The rabbit, happy with its snack, turned and slowly ascended Hulk's hand, nestling into the sleeve of his jacket.

"He'll be just fine with me," Hulk assured confidently. A soldier who was both professional and quirky, Hulk possessed a skill, though occasionally a bit unorthodox, for approaching problems from unexpected angles. He enjoyed tending to bonsais, fancied himself a part-time chef and apparently, now, adopts rabbits.

With the hum of activity winding down, Kane clicked his mic. "Whiskey on me."

Once the team gathered, Kane asked for updates.

Maihuni spoke first. "Med bay and withdrawal routes are prepared."

Pope followed. "Observation points have been stationed and runners are in position."

"Have the two runners been cleared for all our weapons?" Kane asked. The team confirmed they had completed the task.

Pope and Maihuni immediately focused on the rabbit when they saw it. Maihuni's tone softened, shifting from professional to affectionate, as she couldn't resist gently petting the bunny between its ears.

"Just look at you, you little ball of cuteness," she murmured, her tough demeanour melting for a moment. "Aren't you just the cutest little thing?"

Pope inquired, "What's its name?"

Kane couldn't believe what he was witnessing.

Hulk replied, "Don't know. Haven't thought about that yet."

"Can we please focus and forget about the frickin' fur ball?" Kane interjected, his tone clarifying that his question was more of a command.

Hulk straightened up. "Tenders are ready and all non-essential personnel are in their quarters, ready to move. Except for her, she wants to stay with us." They all turned to see Rayna.

Rayna remembered the words of Tom the farmer about the disappearing white rabbit and joked to herself, "Well, it has magically reappeared." Her focus shifted from the rabbit to justifying her decision to stay with the group. She knew this choice would keep her close to danger, but it also increased her chances of rescue and possibly a ticket back to Earth.

As the possible arrival of the enemy drew closer, Rayna's presence became a minor concern for the team. For soldiers, moments of rest were rare during busy days and they sensed an opportunity approaching. Choosing to prioritise downtime instead of dealing with the whereabouts of one civilian who had already been saved once and was becoming a nuisance for the team. Like a Yoko at band practice.

Hulk remained fixated on his newfound obsession, gently stroking the rabbit's fur. The creature, enjoying Hulk's attention, now dozed peacefully in his hand, its ears flopping over one finger as he caressed its back with his thumb. Sensing Hulk as a reliable source of care, the small rabbit rolled over in the giant's palm, inviting a belly rub.

Kane, satisfied that the team was prepared, suggested everyone retreated to their crates while he meticulously reviewed the plan in search of that something that may have been overlooked. He walked through the area, inspecting the viewpoints of both his team and the enemies. Standing before the closed lift door, he imagined himself as the enemy stepping out into the kill zone. His gaze swept left and then right, taking in the long corridors. The enemy would come face-to-face with the hastily constructed barriers of crates and wall panels, just before Kane triggered the snare. Then, it would be too late, they would already be within the kill zone. A classic pincer ambush.

"If they persist, it will come at a high cost. The only logical and survivable direction for them is to retreat into the lift," Kane predicted. "Hopefully, they'll realise that coming by lift a second time won't be worth it." He then organised the lights to be shut down, ensuring that the emergency lighting remained efficient. The minimal illumination worked well, casting enough light on the kill zone while keeping the barricades shrouded in darkness.

Dakota returned with two well-built miners that Kane had requested.

"Thanks, Dakota."

Kane addressed the three men. "I want every man in your platoon to have one of these." He handed over two emergency axes that he had collected from the walls.

"Platoon? What platoon?" Dakota asked.

"The Jingai Volunteers and these two are your first members." Kane smiled.

Dakota shot Kane a puzzled glare, clearly confused. Kane, unsure if the old man had caught his humour, pressed on, explaining that he wanted the men to destroy the operating mechanisms of the first door they passed through, should they be forced to withdraw.

"...and don't do it till you get the signal from the tail end charlie."

"Tail end what?" One miner asked.

Kane pointed to the members of Whiskey. "Tail end charlie will be one of my team members. They will be the last to go through the door and will tell you they are the last. If you seal the door before they exit, we'll be mighty pissed off."

"Understood." All three nodded their heads, fully acknowledging the threat.

Kane was a firm believer in the seven P's: prior preparation and planning prevents piss poor performance. To avoid any poor performance, he walked through the plan with each member of his team until they had internalised his thoughts on the various potential outcomes.

With limited intel on the enemy and their capabilities, Kane considered the possibilities. He wondered if the enemy might have heavy armour, realising that if they did, his team would struggle to stop them.

A well-equipped enemy could unleash devastating firepower, posing fatal consequences for both his team and any civilians caught in the crossfire. Despite these grim scenarios,

Kane felt confident that he had done everything within his power to prepare. He settled down next to Hulk, trying to find some rest before the inevitable clash.

Chapter 21

# CRUMBLING FACADE

With newfound confidence, Rayna approached her temporary protectors near the crates. "If there's anything I can do to help, just let me know," she offered, her voice steady despite her inner turmoil. Kane regarded her thoughtfully, considering her potential as a runner. Someone who could gather people, supplies, or relay messages. He kept her nearby, rather than sending her back to the safety of her quarters like the other non-essential personnel.

Their eyes met once more, and for a fleeting moment, a spark of connection ignited between them. The saboteur who had thrown her well-laid plans for Earth into chaos. Even with her lingering resentment toward Kane, Rayna couldn't shake the unsettling sense of familiarity or the unexpected spark of desire that ignited every time his gaze locked with hers.

She caught a hint of a smile directed her way, though it was hard to tell if it was a smile or a grimace of pain from his wound. Confusion clouded her thoughts. Was it an acknowledgment of their shared moment or just another attempt to hide his discomfort? Yet, that fleeting interaction spurred her to study him more intently, with the same focus she might reserve for a new romantic interest.

As her interest deepened, Rayna took a moment to reassess his height. He wasn't as tall as many Proximian men, standing just shy of one hundred and eighty centimetres, a common stature for their kind. His features were unremarkable, not the type to turn heads, yet there was a rugged charm about him that was captivating. Without hesitation, Rayna seized the opportunity. She ran her fingers through her hair, locating the rebellious strand from before and letting it slip free from beneath her cap. With slow, deliberate grace, she traced it along the nape of her neck, acutely aware of the intensity of Kane's gaze as she twirled the lock around her finger.

Driven by either a sense of duty or something more primal, Rayna knew it was time to take control. "Thank heavens you're protecting us," she murmured, her tone a blend of sincerity and flirtation. Before Kane could respond, the sound of shuffling feet interrupted them, diverting everyone's attention towards Dakota.

"Excuse me, Kane," He interjected, his tone all business. "I just got word that all the unneeded BioStat personnel are in their quarters. The lookouts are in position and a few team leaders are organising any necessary movements."

"Excellent work, Dakota," Kane replied, his focus shifting back to strategy. "This next task is crucial." Kane left Rayna momentarily behind and moved toward Davo, with Dakota in tow. "Listen up Sparky."

Rayna watched them walk back past her, clicking her tongue softly, producing a spark-like sound. "Sparky," She smiled at his nickname. Davo acknowledging her playfulness with a smile.

Kane began, "Hulk will guide you through the withdrawal route," he instructed Dakota. "When the lift descends past us, I want two team leaders from each side of the lift to retrace the same route. Gather all personnel from their quarters on the way to the tenders. Ask Hulk, the one with the bunny, which tenders to use. There's no time for conversations, no luggage. Just move quickly." He then continued with further instructions, his voice growing more serious.

Rayna tried to tune into their conversation, but with Kane's back turned and the hum of the BioStat spinning, she could only catch fragments of what was said. Something about a last resort, but the details eluded her.

When Kane finished and turned around, Rayna's once submissive demeanour had shifted to a captivating gaze that effortlessly pulled him back into her orbit. His thoughts drifted and he found himself mesmerised more than he wanted to admit. There was something about her that was different,, something that stirred unfamiliar emotions within him. He fought to keep his professionalism intact, though the effort was unravelling.

"Maybe you should return to your quarters," Kane suggested, his voice laced with concern, unaware of the impact his words would have.

His unexpected kindness surprised Rayna. Amid the chaos, his injury and her deliberate attempts to distract him, he still showed genuine concern for her safety. She found herself drawn into his web of compassion. Vulnerable and disarmed, her usual predatory instincts faded into the background. For the first time in a long while, Rayna felt comforted, a deep sense of protection enveloping her. This man was unlike any she had ever known.

"Why?" she wondered, struggling to make sense of these unfamiliar emotions. There was no logical reason for the contentment she felt about their connection. Perhaps it was the undeniable truth that it was real. A long-buried giddiness, reminiscent of teenage infatuation, stirred within her. Along with it came a sense of security she hadn't experienced since her father's protective presence in her youth.

"I want to stay here with you," she confessed softly, the words escaping before she could second-guess them.

With that admission, Rayna felt her world tilt. Kane was no longer her adversary. Her deep-seated longing for her ancestral roots on Earth, her convictions, and even her father's misguided desires suddenly seemed insignificant in comparison. Clarity replaced the confusion in her mind.

Everything she'd been searching for might be right in front of her. Here, in this cramped vessel, beside a stack of crates hovering above an isolated rock, she felt no more longing for her distant adventure. Instead, she felt liberated from the fear of exposure and she turned the conversation toward him, wanting to learn more.

The more they talked, the more Rayna felt her old facade crumbling. She knew she couldn't reveal too much, but with every word, she found herself believing less in the lies that once defined her. Her gaze followed Kane as he glanced over at Hulk, who was patting his bunny. "So, are you keeping it or cooking it?" Kane asked.

"I told you, he's our mascot," Hulk replied without looking up, his attention still on his furry companion.

"What's his name?" Rayna asked, feeling less intimidated as she joined the conversation.

Hulk looked up, puzzled. "I don't know," he admitted, as if wondering why everyone thought the mascot needed a name.

# Chapter 22

# THE TATTOO

The security systems were mostly inoperative, with only a few automatically resetting while others sustained irreversible damage. Whatever the invaders did to disable the cameras had left most in a permanent state of dysfunction. Unfortunately, the lift monitors were among the casualties. Replacing the faulty components would be crucial in regaining visibility of anyone approaching via the lift. Davo, aware of Rayna's background as a visualisation designer, saw a potential opportunity. He retrieved another portable workstation from his quarters, hopeful that her expertise might be of help.

"Here, can you help me figure out what's wrong with the lift monitor? It might just be a physical fault," Davo suggested.

Rayna hesitated. "I don't really know much about electronics. I just design parts," she admitted.

"That's fine," Davo reassured her. "I'm still trying to diagnose the issue, but if it's a simple part failure, you could print a replacement."

Despite her lack of confidence in her ability to contribute, Rayna appreciated the distraction from the surrounding chaos. Once Davo logged her into the system, she started searching for the issue. Her attention alternated between the screen and nervously biting her fingernails.

On the other side of the lift, Maihuni's eyes lit up, gleaming with excitement as if she had just struck gold. Unable to contain herself, she leaned over and slapped Pope's shoulder, her grin spreading wide.

"I remember now," she declared, then quickly reached for her coms. "Boss, we need to talk."

"Meet me by the lift," came the prompt reply and Maihuni and Pope headed that way. Even though Rayna was busy working with Davo, Maihuni turned her back to conceal the conversation.

"Something feels off about her," she whispered.

"Go on," he encouraged.

Maihuni asked quietly, her eyes narrowing. "Did you notice the tattoo on her forearm?"

Kane recalled the tattoo and his expression darkened as his memories also recollected its meaning. Etched in his mind was a planetary symbol of Earth. A symbol like that on a Proximian raised too many questions. Kane's initial disbelief gave way to suspicion as he processed Maihuni's concerns. He asked her to dig deeper.

He called over to Brian, "Can you escort Maihuni, please?" His clearance will give her access to where she needs to go.

Kane stood by the lift, his mind tangled with conflicting emotions. He decided to hold off confronting Rayna until he had more answers. Maihuni returned shortly, sparing them a long wait. Watching the lights above the door, Kane calculated he had a brief window before the lift likely left the second BioStat for the atrium. He thumbed his throat mike and the team's radio channel crackled to life.

"Whiskey on me," he ordered, his voice firm. With a quick look at Hulk, who was standing near Rayna, he added, "And bring her with you."

Rayna stood as Hulk approached, confusion written on her face. "What's going on?" she asked.

Hulk shrugged, tucking his rabbit into his jacket. They made their way toward Kane, where the entire team had gathered in a circle. Rayna hesitantly positioned herself beside Kane, the unease of the gathering evident. He felt torn, not just by the accusations he was about to make, but by the bond he shared with her, a bond he was now on the verge of breaking.

He looked directly towards Rayna. "Some concerns have come up that need addressing," Kane started, his tone measured. "I'll speak first, then you can respond."

"Show everyone your forearm," he said, his voice heavy with implication. Rayna's heart sank and she knew there was no escape. She hesitantly pulled up her sleeve, showing her tattoo of a symbol representing Earth.

Her anxiety spiked and she blurted out a rushed explanation. "I got this tattoo ages ago. It means nothing!" she stammered. Normally, such a reaction would seem out of place, but these weren't normal circumstances.

Kane noted her nervousness. Recognising her attempt to steer the conversation away from any potential wrongdoing. But instead of condemnation, he surprised her. "Thank god!" he exclaimed, throwing her off-balance.

Rayna looked up at him with a mixture of confusion and disbelief.

"Earlier, you said, 'Thank god,'" Kane paused, letting the weight of the evidence settle. "And now we learn your tattoo represents Earth."

"So?" Rayna responded, still grappling with the unexpected turn of events.

"Well, it's strange for someone from Proxima to express gratitude towards any deity and coincidently have a symbol of Earth tattooed on their body. Especially with the recent appearance of our visitors from Earth. It raises suspicion." The possibility of his assumptions being incorrect dwindled.

Sneering slightly, Pope added his thoughts, "Earth and their religious fears, fabricated by men to control other men, are nothing but a sham." It's astounding that people still follow that crap blindly."

Hulk added his own blunt view. "Religious nutters, your kind fear, deceive, and rape for money and power."

Rayna, flustered and defenceless, mumbled, "I'm not bad." She saw her ancestors' misdeeds from a distorted lens, blurring the line between good deeds and deception. Their twisted religion had justified oppression, and now the buried truths of the lineage she once honoured were rising to haunt her.

An uncomfortable silence fell over the group. Rayna had no further defence, so Kane pressed on.

With a cold voice Kane revealed, "I tasked Maihuni with searching your quarters and she discovered this," he held up a Bible, "I believe it belongs to you."

Rayna's pulse quickened. She watched as Pope glared at her, his own preaching words biting. "There are no beliefs in the universe, only survival."

Maihuni interjected, her tone dry. "To be honest, survival might just come down to a bit of luck—numbers aligning, preset fail-safes, and scientific formulas."

Hulk ignored whatever Maihuni said, crossed his arms and locked his gaze on Rayna. "People believe whatever they want, don't they?"

"That's not mine!" Rayna's voice trembled, her attempt at defence unconvincing. She wasn't as religious as they believed.

Kane asked Maihuni. "What did you find inside?"

Maihuni held up a page she had torn from the bible. "All sweet things come in small packages. To my little angel Rayna, love Dad." The inscription was damning. She stepped backwards against the wall.

Pope leaned in, intrigued. "Well, well. Now it gets interesting."

Kane pressed on. "Tell me about the mine."

Rayna's face tensed, "The mine? What about it? He saved me," she pointed towards Hulk.

Kane's eyes narrowed. "But no one else in the tubes were handcuffed. Odd, don't you think?"

Hulk's voice cut in. "Fuck, I knew something wasn't right. That's why they didn't fully secure your hands."

Rayna's heart pounded as the walls seemed to close in. The chill from the cool wall at her back contrasted with the fiery panic spreading within her.

Desperation shattered her resolve. "Yes, yes! Fine, fuck it, yes!"

Silence blanketed the corridor as her confession hung in the air.

Pope recalled a quote from somewhere. "My nemesis - my downfall, if you will - was relationships, and trying to fulfill them."

Rayna had never stopped to consider the consequences. The idea that a single mistake could lead to her undoing had never taken root in her mind. 'Downfall' was a word foreign to her father's training in the art of disguise, an alien concept to her life at home. But now, it echoed around her, undeniable and heavy, making her unsure of the future. Her thoughts grew murky, her senses dulled and her body faltered, all because of the blind obedience she had once given to her father.

Kane's voice brought her back. "How did they know to be at the mine?"

"I don't know," Rayna insisted, her voice shaking. "I didn't know they were coming."

"Bullshit!" Kane snapped. "They knew who you were. They wouldn't just handcuff you by accident."

"I swear," Rayna said, her voice cracking. "I only sent information to my father."

Kane's eyes narrowed, clearly interested in this new angle.

"Sparky, bring me that workstation." Davo felt the tension in the air, stood, and handed the device to Kane. Without a word, Kane gripped Rayna's arm and steered her to a nearby crate. With a quick motion, he opened it and gestured for her to look.

"Log in," Kane ordered. "I want to see what you sent your father."

Her hands trembled as she accessed her account.

"Sparky," Kane asked Davo, "can you pull up her video files?"

Davo hesitated for a moment, with confusion at what was happening, but nodded. "Yeah, I can."

As Rayna faced a barrage of questions, her gaze drifted toward Maihuni, who sat on the next crate, scrolling through the files she now had access to. Rayna desperately thought back to conversations with her father, hoping he wouldn't be involved in this. But deep down, she knew what was coming.

Maihuni handed the workstation back to Kane. "She's had some interesting talks with Daddy."

Kane turned. "Go on."

Maihuni scrolled to a particular message. "Here, Dad's asking about the snakes."

The conversations replayed in her mind, and with them, the last of her resistance faded. She finally understood how deeply her father had manipulated her, exploiting her in ways no father ever should. "I didn't want this!" Rayna burst out, drawing everyone's focus. "He's been obsessed with Earth my entire life. He couldn't let go of the power we had there and now he's dragging us all into his mess. I swear I did not know what he was really planning!"

She looked desperately at Kane. "I didn't know they'd come for me. I didn't know why they captured me, but I think it's because of my father. Please... believe me."

Maihuni glanced up from the screen, a look of satisfaction on her face. "Your father's been asking a lot about rhodium."

Kane raised an eyebrow. "Rhodium? Seriously? We'd go to war over that?"

Maihuni nodded. "He even asked about quantities."

Kane considered this for a moment. "Earth's struggling for resources," he muttered. "So, they're poking around here."

"They can buy it like they always have. Why start a war over it?" Hulk asked.

Rayna watched as Kane's expression hardened, the weight of Earth's decline a constant backdrop to everything happening now. She could see it in his eyes, the desperation to protect Proxima from the same fate.

For eons, Earths been on a self-destructive trajectory. Photosynthetic organisms caused widespread extinction, while the growth of plant life led to devastating ice ages. Ironically, the most potent threat to Earth's balance arises from life itself. Humanity, in its pursuit of dominance and insatiable greed, has inflicted more damage on the planet than any previous force.

As humanity remains tethered to the very resources it depletes, it faces the stark realisation that its reign may be nearing its end. During the Apollo missions, when astronauts first gazed upon Earth from space, they witnessed the fragile veil of atmosphere shielding them from the cold void. But despite this profound awareness, political apathy and the influence of religious dogma have fostered inaction in protecting their planetary home.

Telescopes cast their gaze upon an Earth ravaged by self-inflicted wounds. The reckless depletion of natural resources has poisoned the oceans and tainted the soil. These drastic environmental shifts have stirred volcanic activity, intensifying the crisis. Water reservoirs, drained and polluted, now carrying salts that are flung into the atmosphere during eruptions. Wildfires sweep across continents, filling the air with choking smoke. The once-reliable ozone layer has thinned, leaving a life exposed to the brutal intensity of the sun's rays.

Humanity, like a nomadic parasite, has shown little regard for the damage it wreaks upon its habitat. Its relentless exploitation and pollution have unleashed a punishing climate upon the Earth. Deforestation and the degradation of ocean ecosystems, which once absorbed carbon, have pushed them to the brink, turning them from carbon sinks into carbon sources. The resulting heat is intolerable and radiation has decimated entire species. As key biological systems collapse, human agriculture falters. While life may endure in some form, the era of human dominance on Earth is nearing its demise.

Despite centuries of human expansion and the exploitation of space resources, humanity now faces a harsh reality. In order to survive, humanity must move a significant portion of its population to orbit. The colonisation of Proxima, once a pioneering achievement of international space agencies, has given birth to an autonomous entity. The planet's wealth of resources has attracted the interests of powerful corporations, expanding the reach of capitalist ambitions.

Much like the fallen empires of history, Earth now finds itself dependent on fair trade with its former colony. For centuries, the abundant resources of Earth sustained its people, but Earth's relentless gravity has drawn much of its valuable metals deep into its core, rendering them inaccessible. Even resources extracted from nearby asteroids are finite and Earth now faces their depletion.

For Earth, the relentless gravitational pull presents a formidable challenge, making space exploration and resource transportation increasingly expensive. Earth must urgently reform its approach to survival as it faces a decline, with escalating tariffs on critical resources and rising costs of launching spacecraft threatening to cripple its reliance on space ventures. It is imperative to secure the necessary resources for space endeavors, but the question remains: where will humanity find these vital resources to sustain its future among the stars?

Chapter 23

# BLACK SNAKES

Jingai is just one amongst hundreds of asteroids that orbit within Proxima's vast asteroid belt. What makes it special is its resources. Though its supply of water is small, it's enough to facilitate valuable minerals like gold, nickel, cobalt and platinum. However, what truly sets Jingai apart is its abundance of rhodium.

Rhodium, once a niche element on Earth, has become a critical resource in humanity's expansion into space. Extracted from platinum ore, rhodium's initial use was to reduce nitrogen oxide in catalytic converters, but its importance now spans far beyond that. For centuries, it has been indispensable in space technology, preventing tarnish on spacecraft surfaces, protecting vital components such as engines, solar panels and sensors. It also plays a key role in detecting neutron flux levels within nuclear reactors and is essential in 3D printing replacement parts for space missions.

However, its rarity has driven up its demand and cost, leaving Earth's ability to mass-produce orbiting BioStats financially out of reach. A breakthrough in solar energy has shifted the dynamic. Scientists have discovered a way to harness thermal energy from the sun to convert carbon dioxide and water into high-energy fuels. The catalyst is cerium oxide, and the other ingredient is rhodium. Though this process has existed for decades, only recently have scientists developed a portable version suitable for spacecraft. This being a major step forward in achieving efficient chemical storage of solar energy.

The scarcity of rhodium is now at the heart of a tense struggle. Earth cannot meet demand and Proxima, an off-world colony, has paid exorbitant prices to monopolise rhodium resources. Earth's politicians, desperate to secure supplies, have bent and broken laws that were once untouchable. Desperation fuels audacity.

In the midst of the crisis, a descendant of a farm boy walked into a law firm, holding a winning scratchie. The prize: an asteroid rich in rhodium. What follows is a cascade of unpredictable events that will reshape the political and economic landscape of both Earth and Proxima. Setting off a scramble for control of the rare element, pitting the powers of Earth against the ambitions of the stars.

Mining operations now use a highly efficient mass transfer system that has proven reliable and cost-effective. The system comprises elongated cylindrical containers, reminiscent of Earth's old train carriages, yet these "trains" operate track-free. Nicknamed "Black Snakes," these chains of containers have multiplied because of the surging need for raw minerals.

These containers drift in celestial shunting yards, where they await coupling into meticulously organised chains. Once linked, the snakes glide through space, journeying to their respective mines for replenishment before continuing the cycle of resource extraction that fuels human expansion.

The Universal Enforcement Agency (UEA) governs the Black Snakes, much like a galactic traffic authority. Tasked with maintaining order in the mining transportation sector, the UEA conducts routine inspections to ensure that each snake is properly registered and complies with regulatory standards. Their oversight extends to verifying security measures, a crucial step in defending against piracy, a constant threat that jeopardises corporate profits.

Due to the vastness of space, UEA operations primarily rely on unmanned inspection crafts, which patrol the planetary shunting yards. These crafts monitor the waiting snakes for irregularities and enforce compliance within these crowded hubs. Once the snakes leave orbit and embark on their journey, responsibility shifts to advanced AI-guided systems that steer the massive container chains. Lococrafts are stationed at both ends of a snake, synchronise with the AI to ensure smooth navigation and efficient delivery of their mineral cargo.

For bulk resources like coal, now a valuable imported commodity because of Earth's depleted reserves, snakes can consist of over a hundred interconnected containers. With minimal gravity present on most asteroids, launching these massive container chains into space requires little effort, making it an ideal method for transporting resources across the solar system. Upon reaching Earth, each container detaches from the chain and follows a precise flight path, using chutes to slow its descent before landing safely on the surface.

However, Earth faces a unique challenge in this process. Unlike the low-gravity environments of asteroids, Earth's substantial gravity requires considerable fuel to relaunch empty containers back into orbit. This ongoing financial burden has straincd Earth's logistics, making it increasingly difficult to maintain a competitive edge in the resource trade as its previous off-world colony continues to rise in power and efficiency.

Because of security issues with snakes, self-destruct mechanisms are now required on all containers. This precaution, introduced under anti-piracy legislation, activates if a container deviates significantly from its designated flight path. While there have been occasional attempts to hijack containers bound for Earth, the self-destruct feature has successfully neutralised these threats, eliminating the attackers and discouraging future theft attempts.

Beyond its role in countering piracy, the self-destruct system has also proven critical in safeguarding against potential disasters. In the event of AI malfunctions or power failures, the fail-safe mechanism prevents rogue containers from causing catastrophic damage, offering an additional layer of protection in the increasingly automated and perilous landscape of space resource transport.

/ # Chapter 24

# SORRY

"Let's revisit why your father inquired about the snakes. What exactly did he want to know?" Kane pressed, his tone firm yet patient.

Rayna recalled the conversation. "I remember him asking about the docking locations," she said slowly, piecing the memory together. "We also talked through the layout of the mine," she added, eager to be helpful, her voice picking up pace.

Kane nodded thoughtfully as the new information clicked into place. "That's probably the reason."

"The reason for what?" Pope asked, his curiosity piqued.

"They came in through the snakes," Kane clarified."

"Using the snakes would definitely break a few laws," Pope suggested with a knowing glance.

"And it would require serious planning," Maihuni added.

Hulk muttered, "You wouldn't get me in those tin cans. They'd have had to wear masks the whole time. No air, just layers of dust." He paused, realisation dawning. "That guy I ended? He had black soot all over him."

A jolt ran through everyone as a loud "Beerp!" echoed in the corridor, followed by yellow lights flashing above the lift door. Papa hadn't radioed, so Kane knew it wasn't them. The enemy had full control of the system. The lift descended into the mine, with no sign of stopping. Kane remained composed, but tension surged through the ranks. Civilians who had been wandering aimlessly moments ago now stood frozen, all eyes locked his way.

Seizing the moment, Rayna spoke up, her words cutting through the chaos. Despite not thinking it through, she committed herself. "Let me prove I'm not my father's puppet anymore."

Kane's eyes narrowed. "How do you plan on doing that?"

"I'll stay here." Rayna gestured toward a wall just outside the lift. "When the doors open, I'll convince them you've moved on."

Kane's gaze hardened. "And what's stopping you from jumping into that lift?"

"I won't," she said, her voice firm. I've already told you, I no longer want anything to do with them."

Hulk leaned in close to Kane. "If she tries anything, we can take her out."

Rayna couldn't figure Hulk out. One moment, he wants to dine together, the next, he's casually suggesting they put a bullet in her.

Kane nodded. "And we will," he replied, though inwardly he resolved, "I won't cross that line."

"Even if she tries to make a run for it..." Hulk pressed, "... they might just kill her themselves."

"And if they don't, she could give away our position," Kane reasoned, weighing the risks. "But...if she pulls off the distraction. There's a high chance of a high body count. Also, if she does decide to leap into the lift, it will buy us more time.

Hulk waited for the decision.

"It'll do nothing but improve the results," Kane finally said. "She's the one taking the risk. Let's do it."

With the warning sirens now silenced and the flashing light dimmed, Rayna knew the lift had likely reached the mine. Enemy troops would now be entering the lift and the silent countdown had begun in the minds of those present. Time was slipping away, and despite the looming danger, Rayna's new-found loyalty to Kane held firm. Kane ushered Rayna toward the lift. He avoided her gaze, distancing himself from the emotions that would complicate his judgment. This wasn't just about survival anymore, it was about navigating the fine line between duty and personal attachment. In the past, he had lost friends. Despite knowing the pain of the game, this was something different. He wasn't used to putting someone not in uniform into harm's way. Particularly one that he had feelings for.

With a cold voice, Kane detailed her task. "You're going to press the intercom, tell them we've moved and to hurry."

Rayna complied and pressed the button. She relayed the message with a touch of theatrics, her voice convincing.

Kane watched her closely, noticing the slight tremble of her lips. The Enemy pressed with probing questions, but Rayna deflected each one smoothly. When she finished, she looked up at Kane and whispered, "Sorry," aware that she had crossed a line between them.

Kane said nothing, his silence heavy with meaning. Perhaps he had learned a lesson. Perhaps he was simply trying to hold on to his professionalism. Either way, he couldn't let it happen again.

He led Rayna back to her position by the wall, ensuring she was clear of the firing lines. "When the lift doors open, tell them we've moved to the tenders. Say, 'Come quickly,' and make them believe it. When the shooting starts, hit the ground. Got it?"

Rayna nodded, wiping away tears. "I'll do my best."

"Don't make me regret this," Kane warned, his tone final.

She looked up at him one last time, pleading, "Please, Kane...forgive me."

Kane walked away without responding. He knew answering would be an opening, so he didn't.

Rayna braced herself, her mind racing as the lift drew closer. Her hands trembled and a silent prayer slipped from her lips. "Please, God, protect me." But even as she uttered the words, she felt the hollowness of them. It wasn't God who had led her there, it was her father. His manipulations steered her desires to journey to Earth, forcing her to spy on these people, to betray them. She had placed her trust in him, only to find it was all a lie.

The blaring lift alert echoed through the corridor, demanding attention like a persistent alarm. Rayna's memories spun like a pounding migraine, intensifying her anxiety as she stood, waiting for the lift to rise. Her mind drifted back to flickering childhood memories, like glimmers of wax rising and falling within a glass vessel.

She recalled the laughter, the simple joys and the moments when her family felt whole, when life was still untouched by the shadows that now loomed over her. Suddenly, clarity pierced through the fog of her mind. Her father had manipulated her and she was done with allowing herself to be his pawn. She now wished to return to Proxima, not Earth, to break free from his grip and reclaim her life.

Kane peered over the crosshairs of his weapon, quietly admiring the courage that had led her to this point, despite the peril that surrounded them. Her strength stirred something within him, causing him to question if he would make the same choices if their roles were reversed. He wrestled with doubts about of his morality, understanding now that good and evil weren't as simple as he once believed. Perspectives shape your actions and what one saw as evil could merely be the path another had no choice but to take.

Sensing Kane's gaze, Rayna straightened her posture and collected her thoughts. Her fingers brushed through her hair as she removed her cap, letting her locks flow free for a moment before replacing the cap. She saw it as a subtle rebellion, a minor act of resistance against her father's control. She glanced back at Kane one last time, their eyes locking for just a moment. At first, Kane remained still, but when their eyes finally met, he offered a subtle, deliberate wink, a quiet reassurance amid the chaos to come.

Reflecting on what he might say to her, Kane remained silent. His duty demanded unwavering resolve. Rayna confronted her fears and made the decision that would define her character. Deep down, Kane knew he had made the right call. Despite the weight of his responsibilities, he trusted in his ability to navigate the challenges ahead, reaffirming his commitment towards being a devoted SERT leader.

As though attuned to Kane's inner turmoil, Rayna's tense expression softened, giving way to calm acceptance. She was ready to embrace her past mistakes and resign herself to whatever fate awaited her. With a defiant gesture, she removed her signature snoopy cap, allowing her hair to cascade freely. The symbol of liberation from both the mine's oppressive grip and the burden of her past. Her eyes found the dim figures of Pope and Maihuni behind their barricade.She turned her head toward the opposite crates, her eyes drifting to Kane. Her gaze softened with affection, lingering a moment longer than necessary.

Unbeknownst to Rayna, this fleeting gesture marked one of the rare moments Kane allowed tenderness to surface. Despite his inner turmoil over her betrayal, she had embraced it as a path to redemption. That fear that once held her was now gone, replaced by a newfound confidence. She shifted her posture, easing the pressure of her bra straps on her shoulders and focused her gaze on the nearby wall sign that read 'EXITING TRAFFIC.' Though insignificant, the sign captivated her, its surface catching the light with the interplay of shadows cast by the pipes above.

Lost in thought, she wiped a bead of sweat from her cheek just as the warning lights signaled the lift's approach. Startled back to reality, she cast one last glance at the barricade only to find Kane and the others had mostly disappeared from view. Her focus centered on the lift doors that were soon to open, her mind racing with the decisions she would soon have to make.

As the doors slid apart, a flood of light forced Rayna to squint. She recalled Kane's advice to close one eye, doing so until her vision adjusted. But what she saw froze her in place. A wall of shields advancing, blocking the doors from closing. Behind the shields, soldiers crouched on guard. Slowly, the shields parted at the centre, revealing a commanding figure of an intimidating stature.

Team Whiskey remained vigilant, their eyes trained on the kill zone. They waited for Kane's signal to engage, but his attention was drawn to Rayna, knowing her crucial role in the mission's success. He noted her nervous gaze toward the light and sensed her apprehension.

The enemy formed an arc with their shields around the open lift doors, ready for a fight. Kane held his position, poised to fire. They remained out of sight, and fortunately, their shields kept them from spotting the ambush. With steady patience, he waited, confident in his concealed vantage, silently urging them forward. "Just a little further."

Kane watched with rising tension as the enemy commander approached Rayna, his form largely obscured by the shields of his soldiers. Bathed in the light from the lift, Rayna stood like a performer stepping onto the stage for the first time. A solitary figure awaiting judgment from her new audience. She angled her body to cast her face into the shadow of the authority figure as he stepped out. His towering presence, marked by broad shoulders and an aura of confidence, demanded attention. The light haloed around him, obscuring details, but Rayna could still make out his high cheekbones and piercing, distrustful eyes. Even in the partial glow, his features struck her as those of a finely sculpted masterpiece.

Rayna's voice broke the tension. "Thank goodness you're here. I'll take you to where they went." She opened her hands in an inviting gesture, stepping into the role she needed to play to survive.

He paused, studying her invitation with a caution. Then, to his right, a glint of reflection off a structural beam caught his attention. His instincts, finely tuned and heightened, were honed by years of command. He sniffed the air like a predator sensing a trap. His ears prickled with unease, sensing that something was wrong, perhaps a trap.

In the shadows, Hulk quickly snatched up the rabbit and tucked it back into his jacket.

Time seemed to stretch as Rayna's eyes met the commander's angered gaze, the coldness behind them freezing the moment. Suspicion flickered and sharpened into something more dangerous, hardening his resolve. Slowly, almost methodically, his hand secured his rifle against his chest mount. The world around them blurred, each movement magnified. His fingers wrapped around the pistol's grip in a slow, fluid motion, every movement precise and controlled. With precision, he released and raised the weapon, his confidence unshakable, as though he had orchestrated this moment long before either of them realised it.

In one fluid motion, he aimed and pulled the trigger. Rayna, untrained and paralysed by fear, had no instinct to flee. Even if she had, it would not have saved her. The energy blast struck her temple with brutal precision, tearing through her skull. The force snapped her head back violently before it jerked forward again. Her body collapsed to the ground, lifeless and still.

Kane's reflexes were lightning-fast. He unleashed a relentless stream of laser fire, each pulse slicing through the air with a searing intensity. His team quickly falling into rhythm as they joined the ambush. The enemy commander took cover behind his soldiers' shields. Their weapons were ineffective against the impenetrable barrier of shields, resulting in the corridor erupting into a chaotic display of lights and shadows. The intense gunfire created a mesmerising spectacle, effectively suppressing the enemy and granting them valuable time.

Rayna was no longer everyone's focus. Every muscle in her body relinquished its hold, allowing her lifeless form to crumple like a collapsed accordion. Twisted on the floor, her eyes vacant and lifeless as she lay utterly motionless. No longer the fiery pocket rocket in her father's schemes, but a memory left behind in the unfolding struggle.

The only visible targets were huddled behind their transparent shields, their silhouettes mirrored in the reflections. The enemy, as expected, shielded themselves from the constant barrage of gunfire, desperately avoiding any vulnerable exposure, fully aware that the chance to return fire wasn't out of the question.

With the ambush executed, it became clear that the conflict's intensity was unsustainable. Kane feared the situation may escalate with severe consequences. Recognising his team's firepower limitations, he began considering the need for a tactical withdrawal. Then the enemy gave way. Knowing that returning fire would lead to casualties, they retreated into the safety of the elevator. Kane signalled his team to cease firing as the lift doors slid shut. The last blasts gave way to the clanking of reloads and battery cells clicking back onto their chest plates. Then there was silence. Maintaining their professionalism, they remained ready, waiting for further instructions. The silence was broken by the 'Beerp, beerp!' of the lift as it began its descent to the mine.

Chapter 25

# THE SCHMELLER

Kane's gaze locked onto Rayna, her eyes wide and pleading, silently begging for intervention. He stood still, waiting for any sign of life, though he knew deep down there would be none. Maihuni stepped forward, rolling the body over and checking for a pulse. It was a pointless gesture, yet one made of respect. Rayna's face, framed by singed, messy hair, bore the brutal marks of her fate. Her forehead, a tangled mess of charred strands, hid a fractured wound that would haunt Kane for nights to come. He exhaled sharply, the breath escaping through his nostrils and he shifted his attention back to the living. Kane's eyes scanned rapidly over their dwindling options.

Davo rushed to his side, driven by a fierce protective instinct. His eyes darted toward Rayna, desperate to do something, anything. But Kane's voice cut through the tension like a knife. "Don't!" He warned.

Davo paused, his expression darkened. Despite Rayna's earlier confessions complicating things, Kane was surprised Davo relented so easily. What came next, however, was even more unexpected.

Davo muttered, his tone bitter. "When her god was handing out brains, she must've mistaken them for smoothies, because she ordered an extra-thick one,"

Kane admired Davo's ability to maintain composure under pressure. People handled death differently, especially when it hit close to home. Still, Kane couldn't help but feel that Davo's reaction was a bit too cold.

Then Davo's focus shifted, his voice hardening. "That prick needs to die."

For soldiers, taking another life was sometimes the ultimate duty. But the enemy commander had crossed a line. Rayna's death had been unnecessary, an act devoid of professionalism. It set the tone for what was to come. That man has made himself their primary target, a symbol of their hatred, a focus of their need for revenge. Warfare could be swift and brutal, but vengeance had its own rewards.

Kane placed a steady hand on Davo's shoulder. "I can't promise anything, Sparky, but you'll be the first to know when we catch him."

Pope chimed in. "The Schmeller. That son of a bitch will get what's coming."

"The Schmeller?" Davo asked, confused.

"Yeah. Didn't you see that arrogant, aristocratic nose? I'm surprised he got it to fit in the lift."

Without proper logistics, any plan was destined to fail. The enemy knew Whiskey was light on firepower and their only hope was to return to the Centaurus, where they would regroup and dictate the battlefield. It became obvious that defending the BioStat was futile. Kane radioed Papa One, stationed above in the BioStat, granting clearance for the tenders to depart. He signalled Whiskey to begin their withdrawal. Hulk led the way, lumbering down the corridor as the retreat unfolded. On the opposite side, Maihuni mirrored his movements. Kane and Pope held their ground, waiting to ensure everyone had cleared the area before following. As they reached the first doorway, Kane glanced at the volunteer standing nearby, gripping an emergency axe.

Kane's voice echoed through the coms, sharp and unwavering, "I'm the last one. Destroy it!"

Without hesitation, the miner swung his heavy tool, smashing the control panel to pieces. Sparks flew as the electronic components crumbled under the blow. They moved on, withdrawing through the BioStat until they entered the vast biome. The sudden rush of humidity clung to Kane's skin, filling his lungs with the thick, damp air. It always struck him how alive the artificial climates felt.

Proximians seldom experienced the subtleties of natural weather. They don't understand the satisfying crunch of autumn leaves underfoot or the crisp bite of morning air on an early commute. They couldn't imagine the ache of yearning for the warmth of a blanket as winter's chill settled in, nor the scorching heat of summer afternoons spent body surfing with family at the beach. To them, these simple pleasures were little more than stories from a distant, mythical world. While BioStat's biomes sometimes recreated these sensations, the connection to a real earth environment was lost on those who had never experienced it.

They passed the trickle of fishponds, the pungent odour of animal cages and scattered rabbit hutches that littered the area. Kane recalled the ambush. He suspected the rabbit was likely the trigger, but there was no time to think about that now. Ahead, Hulk organised the final loading of the tender. Kane positioned himself to guard the embarkation point and glanced towards Davo, who was monitoring his device. "Anything?"

"Nothing," Davo said, fingers flying over the keyboard. "Wait, scratch that. They've tripped the lift sensor."

Kane stiffened. "How many?"

"Five," Davo confirmed after a quick scan of the footage.

Kane did the math. Five inside the BioStat. One dead in the mine, which left possibly two unaccounted for. That's if there's only two teams.

"Tell Hulk to speed it up. We've got company," Kane ordered.

Hulk helped people into the tender, then Dakota herded them further inside, shifting a few from the hatch to make space for Kane, Davo, Hulk and himself.

Standing guard alone always weighed on Kane. He scanned the quiet farm area, where the only sounds were the hum of insects and the soft clucks of chickens. The peace felt unnatural, even eerie. His mind wandered, as it often did in these moments of stillness. He couldn't stop thinking about Rayna. Had falling for her been a mistake? Maybe it was a trap from the start. Had she played him? No matter how much he doubted, the memory of her beauty clung to him. It had felt like fate, fleeting but powerful. Now, the desire to confront her killer burnt inside.

Davo returned, tapping the pipes above Kane's head. "Time to go," he said.

That was all Kane needed. One man with a pistol had no chance against five armed with rifles. He glanced into the tender. The passengers crammed themselves so tightly that there was hardly any space left.

Panicked eyes followed Kane as he squeezed into the last open spot on the floor. Parents were struggling to quiet their children, unable to make them understand the urgency of the situation. Their entire young lives had always balanced on the edge of survival and this moment was no different.

Tom, the farmer, started grumbling about leaving his animals behind. Kane shot him a look of pure disbelief, the absurdity of the complaint clear on his face. His frustration was palpable, even to those watching from within the tender. He tapped Dakota on the knee, giving him a nod of thanks for playing the impromptu role of flight attendant, organising everyone for departure.

As the pilot ran through his pre-flight checks, Kane sealed the hatch. Relief flickered across his face, safety now within reach. He plugged the tender communications cable into his headset. "Trashie, give 'em the countdown."

The pilot complied, breaking radio silence. "This is Six-Three-Charlie, separation in three, over."

Acknowledgments crackled in from the other tenders. No one wanted to wait. The sooner they got away from the BioStat, the lower the chance of being caught in a firefight.

"Rightio, Sparky, shut 'em down," Kane muttered.

As planned, this was Davo's cue to deliver his parting gift. His fingers flew over his device, executing commands that sealed every door in the BioStat. No one inside was leaving without some serious hacking.

He frowned. "We've hit a glitch with the last tenders. The system bypassed their lockdown and flagged them for emergency evacuations."

Kane waved it off. "Doesn't matter. The tenders won't change the outcome." His mind eased, knowing everyone was safe. "Hand the device over to Dakota."

Dakota took the device, logging in and navigating through the security protocols to access the BioStat's safety system. He braced himself for a task he never thought he'd be responsible for. "I'm ready," he said, his voice steady.

He waited for Kane's signal.

"Just a little longer." Kane watched the monitor, tracking the tenders distance as it moved from the BioStat.

Moments later, Kane gave the order. "Okay, that's far enough. Do it."

Dakota pressed the button, triggering the release. The Bio-Stat's fail-safe rockets ignited and the shafts' mechanisms whirred to life. Rotating the entire station on its shaft and unlocking it from the asteroid's hold. Slowly, the station drifted away from the asteroid, carrying with it at least five enemy soldiers.

Brian smirked, his voice barely a whisper. "You're not keeping what's ours," he muttered, as though the invaders might somehow hear him. He knew that once the chaos was behind them, it would only be a matter of time before the Centaurus' tenders restored the BioStat.

The tender made distance out of the shadow of Jingai and towards the Centaurus with the others. There was a noticeable shift in the mood inside the cabin. The tension slowly ebbed, replaced by a sense of cautious relief.

"I'd trade anything to be on the Yuri Gagarin right now," Dakota mused aloud.

"The Yuri what? Never heard of it," Davo replied, looking genuinely puzzled.

"Yeah, the Yuri Alekseyevich Gagarin," Dakota continued. "The original ship Earth sent to colonise Proxima. It took decades to get there. It's probably still floating around Proxima's airspace as a farm."

Hulk, grinning, joined in. "Right and we'd be on that ship right now, sipping beers, eating good food, watching the kids play. No worries, no problems, not a care in the universe."

Davo raised an eyebrow, his interest only slightly stirred. "What the hell are you two on about?" He cared little for history, but the conversation was a decent distraction.

Kane laughed, shaking his head at the absurdity of the idea. "Yeah, sure, because you'd rather be lounging on some ancient relic than out here, in the thick of it, with me?"

"Not the point," Hulk said, chuckling. "Besides, I'd have a bunch of kids, each trained in the art of silence, of course."

"And we'd train them to pour our mead," Dakota chimed in with a wink. "Bet you've never trained your team to do that, Kane."

"Cheeky prick," Kane thought with a smirk. "Now that he's safe, he thinks he can bullshit with the rest of us." Still, he knew better than to disrespect his elders.

Noticing Davo's puzzled look, Dakota seized the moment. "The Yuri Gagarin was one of the earliest generation ships sent from Earth to colonise Proxima. Space travel was still primitive back then, so these ships crawled through the stars. They figured it would take about 120 years to reach their destination. Generations would be born, live and die on board, which is why they called it a 'generation ship.'"

"Got it." Davo clarified.

"Anyone would think Pope was here with us." Kane pointed out Dakota's sudden change in confidence and feeling the need to share it.

Dakota continued, rubbing his chin as if savouring the weight of his own explanation. "So, there we were, cruising along one day and we passed the Gagarin. The funny thing is, we've already settled Proxima and here come our ancestors, still making their way there."

Hulk added in a wistful tone. "Just imagine, no corporations breathing down our necks, no political crap. Just kickin' back, sippin' mead and watching the stars drift by."

Kane interrupted, "Sorry to burst your bubble, but given the choice between getting old with a wife and a bunch of unruly kids or being out here with me, I think I know what you'd pick."

"You're missing the point, Kane," Hulk replied with a grin. "Every one of those kids would probably have a different mum."

Kane just shook his head as the laughter echoed through the ship.

# Chapter 26

# GRITTED TEETH

Docking with the Centaurus was almost complete. With the slow, rhythmic thud of metal meeting metal, the walls of the tender groaned and equalised. The larger vessel exerted a subtle rise in air pressure against the tender's hull, causing the floor plating to tremble gently. Echoing creaks and knocks filled the air as the docking clamps engaged. For those unfamiliar with the process, the experience was both awe-inspiring and disconcerting. Wide-eyed stares followed the shifting shadows cast by the lights, while the soft, almost imperceptible pop of their ears indicated the last adjustments in pressure. The docking sequence was a precise ballet of technology and engineering, every sound and vibration a reminder of the vast, interconnected machinery that held their world together.

Kane and Hulk were the first to disembark, undergoing delousing procedures before being escorted to a designated meeting room. Pope and Maihuni followed shortly from their tender. Standard protocol dictated that the team are to be separated and debriefed to extract unbiased intelligence. Kane, as the team commander, anticipated a more rigorous debriefing than his team. Before they're separation, he issued instructions. "Once we get this shit out of the way, clean your weapons and get your full kit ready. There's no more falling back," he ordered, before adding, "Rest if you can, but be ready to move at a moment's notice."

Hulk's companion stayed concealed in the cramped storage compartment of the tender, tucked away beneath a jumble of supplies.

The fewer people who knew about this furry stowaway dodging quarantine, the better. With its twitching nose and curious, soft eyes, the little creature had stayed out of sight, but its presence added a subtle layer of tension for Hulk.

Once in the debriefing room, Whiskey's responses were sharp and concise. The room was a blur of faces and voices. Each question asked carried the weight of necessity. Each response meant Whiskey were closer to addressing more pressing matters and ultimately, an opportunity to rest.

Once dismissed, they dispersed to the armoury, collecting Hulk's mascot on the way. The ritual was automatic, clear the SIPLATs, disassemble, clean and reassemble every part. Then, they submitted their batteries to the armourer for recharging. Their SIPLATs would now be the smallest weapon they would carry.

"It's good to be back," Pope said with a satisfied grin, feeling the space settle around him like an old friend.

The armoury, a second home to the team, provided more comfort than their quarters. Within its walls were not just weapons and equipment, but the personal items soldiers clung to: a deck of cards, a table, a fridge. It was a place of shared memories, a blend of power, mateship and yarns.

"Time to break out the heavy artillery." Hulk was excited. With their payload beefed up, so too were their spirits.

They signed out their kit, which was meticulously sorted and stacked for swift retrieval, when the call came.

Afterwards, Hulk, Pope and Maihuni headed for their quarters and a much-needed shower. Hulk chatted to his rabbit like it was a green recruit, welcoming it to this hideaway. "Welcome to our wank booth," he muttered as it nestled under his bed among dirty clothes and food wrappers. Hulk's post-mission routine was ingrained: secure the door, empty pockets, boots and gear off and dive into a hot shower. The transition from the intensity of work to downtime was often jarring, but the warmth helped ease him into it. After the shower, he indulged in simple comforts—music, messages from home and plants that needed watering. Each member has their own habits. Pope's is reading. Maihuni's is the gym. Kane's interests remained a mystery to most, although his fondness for mead was well known. Each found solace in the repetition of their favourite downtime distractions.

Kane returned to the armoury after completing his debrief and battle report. The XO was waiting for him, ready with the details of the upcoming meeting. Kane's team had already laid out his gear and he took a moment to arrange it to his liking before retiring for a shower and then a meal. There was no room for a lengthy break. As always, he reminded himself, "It's part of the job."

He grit his teeth in pain as the heat from his shower irritated his wound. But he stood still, letting the pain dull to a throb, a small price to pay for the comfort of heated water. Deciding an extended shower was necessary, he adjusted his plans, opting to eat on the go, followed by a quick visit to the med bay.

As the warm recycled water flowed down his back, his thoughts wandered to Rayna. He stood there, pensive, a man acquainted with the ache of absence. He'd loved and lost before, but her absence now haunted him. His inner turmoil was his alone to navigate. He wouldn't share this burden with anyone, especially not his team. Weaknesses make you a target and Kane refused to let anyone know that a woman could be the catalyst for his downfall.

With a fresh uniform, Kane grabbed a quick bite before heading to the med bay. He found himself distracted by posters of idyllic landscapes adorning the walls. They depicted scenes of a brighter, more expansive existence. Landscapes of a planet that was now of his enemies. The doctor noticed his gaze. "Looks like the posters are doing their job," he remarked, breaking Kane's trance and pulling him back to reality. Kane's mind wandered to thoughts of his future once again, but the uncertainty of the current situation made it easier for him to snap back into focus.

The routine medical checks began. Radiation levels, muscle and joint mobility, eye function and then the doctor focused on Kane's stab wound.

"Maihuni has already restocked her supplies," the doctor pre-empted. "She's not bad at patching people up. For a grunt, that is." His tone was dry but respectful and after applying a sterilised patch, he advised Kane to take it easy.

"Easier said than done." Kane replied with a smirk.

Chapter 27

# DEFEAT A SNAKE

As was their habit, Kane and the two other SERT leaders arrived early at the command centre meeting room. The events that were taking place were rare, but big gatherings were common. The room, although not spacious, could accommodate the complete tactical command structure and most of the support staff for the imminent operation. Kane viewed the monitors displaying the familiar maps and images: Jingai, the BioStat drifting in space and now, the ominous silhouette of the Earthen liner.

The captain strode into the room, tugging at his tunic with rushed movements. It was clear that personal appearance had taken a backseat during this period of diplomacy. With even less time for the usual courtesies, his demeanour was more hurried than ever. "Jingai used to be in the middle of nowhere. Now it's in the middle of everywhere," he remarked, his tone sharp as he jumped straight into the briefing. "Alright, Jes, let's get started."

Jesse, the intel officer, nodded and launched into a concise rundown of the data flashing on the monitors. "Most of you know of our teams' commendable work earlier. They released the BioStat from the asteroid and safely evacuated the inhabitants to the Centaurus. He referred to the live feeds. "Intelligence confirms our adversary is from the Earthen liner Tyson, which is currently stationary on the far side of Jingai." Jesse pointed to the monitor, displaying coordinates. "Blueprints show it's configured much like the Centaurus. The XO will provide updates on any deviations shortly.

The room was quiet, attendees taking notes as the intel officer continued. "We suspect the enemy infiltrated the asteroid tubes by hitching a ride in the snakes. All snake activity near Jingai have been paused for now. We are still determining if any enemy forces remain on the asteroid, but we believe there are possibly seven individuals trapped in the BioStat. They're fully armed, including shields. Tyson, the enemy liner, has just sent a tender to the BioStat, suggesting that they will retrieve their teams. Enemy SERT numbers aboard the Tyson are unknown."

He paused, scanning the room. "Any questions?"

A tender pilot raised his hand. "Can you clarify their motives, sir?"

"Earth is claiming ownership of the mine, lodging a formal claim on the asteroid. Settling these disputes is a long process, made worse by diplomatic complexities. Proxima has already shown Earth's claim is legally flimsy." Jesse explained.

"Then why the attack?" the pilot pressed, echoing the confusion in the room.

Jesse sighed. "We believe Earth likely expects a negative ruling. It's about consumerism. Our assessment shows they're desperate for rhodium and Jingai happens to be one of the nearest richest sources."

"Starting a war over rhodium?" someone murmured in disbelief. "There's plenty of it out there."

The captain cut in. "Speculation suggests Earth is in deeper trouble than we know. Resource costs for orbital operations are steep, especially with their gravity issues. Their costs are higher, their demand immediate. Desperation drives desperate actions."

"Or perhaps this show of force is meant to sway the courts," Jesse added with a shrug.

The captain nodded. "Thanks, Jes. XO, you're up."

The XO stepped forward, laser pointer in hand, drawing attention to an image of the Tyson. A man of power which is echoed in his voice and stance. "To kill a snake, you cut off its head," he declared.

Kane felt a jolt of anticipation surge through him at the XO's words.

"Ladies and gentlemen, we are taking the fight to them," the XO continued. "Our intel has pinpointed the enemy command — here," he said, directing their attention to the aft superstructure on the blueprints.

Pens scratched furiously across notepads as everyone absorbed the intel.

"The exact number of teams aboard the Tyson is unclear," the XO noted, "but we estimate between three to five teams. Prepare accordingly."

A murmur of concern rippled through the room at the idea of launching an attack against uncertain numbers. Standard military doctrine favours a three-to-one ratio for the attacker, making these even odds highly unfavourable. With the possibility of facing superior numbers from the defenders, compounded by the prospect of the ship's crew joining the fight, the mission felt perilous.

Sensing the unease that hung thick in the air, the XO straightened, his gaze sweeping across the room. The low hum of the ship's engines underscored the tension as each team member stared at the tactical display, their faces illuminated by its cold blue glow.

"I understand these odds aren't ideal," he began, his voice cutting through the heavy silence, "but who's counting when we have the best?" His words delivered with a wry smile.

The SERT leaders, seasoned veterans of countless high-risk operations, met his gaze. They were a hardened crew, accustomed to navigating through chaos with the precision of well-oiled machinery. A few exchanged knowing looks, silently recalling the near-misses and impossible odds that had forged them into an unbreakable unit.

He continued, "We don't have the luxury of hesitation, not here. Not now. I know what each of you is capable of and I trust you to follow through. No hesitation, no complaints. We do our job and we do it better than anyone else in this galaxy."

The rhythm was a familiar one: facing odds, overcoming obstacles by any means necessary. The SERT leaders in the room were driven by a singular purpose, to complete their mission, whatever it takes.

"As you're aware, regulations prohibit us from initiating an assault on a ship's exterior," the XO continued. "But it's foolish to assume the enemy will play by the rules, given their earlier record of dishonourable conduct. "If the tenders are attacked during the initial transit, the mission will be called off, and all tenders are to return to the Centaurus. We need to uphold the appearance of legality. Any actions that fall outside the rules could sway public opinion against us, strengthening Earth's position."

Quebec's team leader chimed in, reflecting on past training encounters with the Tyson during his station at YX49, the massive asteroid near Uranus. "The ship's executives were predictable and always played to their strengths."

"Excellent!" the XO affirmed. "That's precisely what we're counting on." He turned to the SERT leaders. "Once aboard, you'll level the playing field by controlling the battleground."

Quebec's team leader leaned over to Kane, his voice low with a touch of bravado, engaging in the familiar banter that soldiers often share. "I took my first ever bath on the Tyson, won't ever forget it. Shared it with two of the ship's female junior staff," he whispered, though this wasn't one of those moments of idle chatter.

The XO continued with a detailed situation report, assigning personnel to each SERT leader. Engineers from both the ship and BioStat's crew were designated to join the teams, along with pilots and two volunteers per tender. Support tenders would accompany, equipped with portable medical modules and supplies.

Turning to the medical officer from BioStat's crew, the XO said, "Alright, Doc, your quiet tour ends here. Prep a surgical team for the forward structure and have two teams on tenders ready to go."

"Yes, sir," the medical officer replied.

The XO then addressed the chief steward. "If we capture prisoners, they are to be treated well and secured once aboard the Centaurus. Ensure you have a module ready for that purpose."

"Understood, sir," the chief steward acknowledged.

The teams received orders. The first phase involved SERT's Papa and Quebec inserting into the Tyson's forward superstructure. Their mission is to create the deception of a full assault to draw enemy forces away from the other superstructures. Forcing the enemy to engage on their teams. The strategy aimed to bait the defenders into committing large numbers to overpower the breach. Then they are to establish a moving defensive stance that is to keep continual contact with the enemy.

Once this initial assault is underway, SERT Whiskey's tender will pretend to reinforce the breach, only to suddenly change course towards the enemy's command centre in the aft superstructure. If successful, the enemy would have already committed themselves, hindering their ability to respond effectively elsewhere.

"Whiskey, you have two objectives: Block the enemy from returning through the ship's spine and seizing the command centre, stripping the captain of command." He stated as though it was a picnic.

Kane carefully deliberated on his objectives, recording them below the roster of personnel assigned to his team. A signal from the XO's technical team flashed on Kane's tablet, providing him with the Tyson's blueprints and key objective locations. He knew not to look until the XO finished speaking.

Finishing his orders, the XO handed control back to the captain, ready to set the pre-mission preparations into motion.

"Thank you, XO," the captain acknowledged before turning his attention to the room. "Capturing the Tyson won't single-handedly halt Earth's operations, but it will serve as a deterrent against further actions at Jingai while we wait for more support." He paused, scanning the group. "Now, I need to finalise orders with those staying aboard the Centaurus." He gestured toward the exit. "So, if you're going on this picnic, go get your baskets ready." Turning to the mess officer and added, "Gordon, make sure the security detachments are well-fed."

"They always are, skipper," came the confident reply.

Those about to put themselves in immediate danger filed out. Kane messaged his team and support staff, informing them to meet at their armoury. Davo arrived shortly after Kane. "Please tell me you're not sending me back to BioStat?" he joked.

Kane fixed him with a serious look. "Sparky, you fucked up," Kane said bluntly.

Davo frowned. "What do you mean?"

"You did such an outstanding job on the BioStat that you've now been attached to us for our next adventure," Kane replied, omitting the part where he had specifically requested Davo's presence.

"You're kidding. What adventure?" Davo asked, clearly in disbelief.

"We're paying a visit to the Earthlings. Stick close and when we're done, see the armourer to get suited up." Kane said.

Davo's confusion shifted to a knowing smile, seeing an opportunity for payback. Kane wasn't the only one haunted by Rayna's death. Hulk gave him an encouraging pat on the back. Ever since discovering their mutual love for food, he'd taken a liking to him.

Kane's team had expanded to eight, including Davo, two volunteers and the tender pilot. Kane addressed the group. "Alright, listen. The skipper's rule is whiskey or nothing, so he's dispatching Whiskey to visit the Tyson." He briefed them on Whiskey's critical role in the mission, underscoring the need for precision and determination.

The team prepared their kit and moved to similar sections of the Centaurus' superstructure to familiarise themselves with the ship's layout. They rehearsed breaching manoeuvres, navigated routes to key objectives and practiced operating in full gear. Kane drilled them on hand signals and movement procedures, aiming for a smooth operation, minimising the need to explain the obvious under fire.

Identifying possibilities they would encounter on the Tyson, Kane turned to Davo. "Sparky, we need this door secured from this side. Think you can manage that?"

Davo nodded thoughtfully. "It'll take a while, but I think I can do it."

Kane reconsidered. "Actually, scratch that. Let's use a door brace instead. Just be ready to disable the lock so you can re-secure it.

"Understood," Davo confirmed.

Kane turned to Brodie, a recent addition to the team. "So, Brodie, how does it feel to be a soldier now?"

Brodie hesitated, then answered, "Excited, but also nervous as hell."

Kane chuckled. "That's normal. We all feel that way."

Addressing the other recruit. "Daniel, make sure you're paying attention." Kane never forgot a name once introduced. "Unless we give you other duties, you and Brodie will be on guard detail." This room aboard the Tyson will serve as our holding cell. You're responsible for escorting prisoners here, that's if there are any, and securing the door. Kane demonstrated by releasing the brace strap of the brace he was carrying, gripping both ends firmly. He positioned it between the doorway frames, extended the brace until it jammed tightly across the door. With a quick turn of the locking latch, two suction pads activated, gripping the door's surface securely. As an added measure, two sliding latches automatically extended along the brace, locking into any recesses in the door. To prove its effectiveness, Kane leaned his full body weight against the brace, testing its hold. Satisfied, he released the lock and retracted the brace back to its original size.

"Got it," they both replied.

"Now practice it." Kane ordered.

Kane stressed the importance of vigilance, highlighting the risks of failed prisoner security and the potential dangers and repercussions of complacency. "Watch each other's backs. I don't want to be the one telling your families something went wrong because of your screw-up."

They moved on to practice evacuating casualties, both friendly and enemy, emphasising how teamwork was critical in protecting each other. "We stick together and we all get back to our ship together. Understood?" Kane's tone was unusually solemn.

"Alright, everyone, grab some grub, take a crap and finish getting your gear sorted. We load the tender at nine bells."

Daniel laughed nervously. "How am I supposed to shit under this kind of pressure?"

Hulk grinned, glancing at Daniel. "Welcome to SERT. Just be glad he's not specifying what kind of shit you need to take."

Chapter 28

# HURRY UP AND WAIT

The Centaurus pilots maneuvered their tenders, shuffling containers from the storage zone surrounding the ship. Sticking closely to a meticulous schedule that mapped out every cargo's location, they moved a few containers at a time, tethering them temporarily to long lines extending from the Centaurus. Reshuffle stacks, they reach the breaching and medical modules. The designers primarily intended the breaching modules for rescue and recovery missions, not military insertions. This time, they will load the breaching modules with the necessary gear and supplies that will pack a punch.

Once securing the modules to their tender's cupolas, the pilots began returning to their docks, where their respective teams awaited boarding. Other tenders promptly began attaching containers back onto the Centaurus, reshaping it into a dense, interstellar warehouse. Six-Three-Charlie's pilot aligned the breaching module with the docking bay, triggering a series of alarms as the ships synchronised.

"Docking complete," announced a monotone computer voice.

An assortment of gear and supplies piled against the wall at the docking port's entrance. The supplies included weapons, medical kits, power units, shields, rations and a hidden stash of centaur mead. Kane pulled two levers. Feeling the release, he opened the docking door and then spun the wheel, releasing the module's hatch. He turned and addressed his team as he grabbed a couple of equipment bags. "The captain wavered delousing for this trip. Figuring he's happy to share the bugs and germs we have with the Earthen fuckers."

Descending from his cupola, the pilot joined the crew in loading the module. "And again, welcome aboard Six-Three-Charlie," he greeted. "Let's make this run as memorable as the last."

"Ah, Trashy, it's you!" Pope laughed, recalling their last encounter. "Let's hope your galley has improved. We haven't forgotten the culinary disaster you served on our last voyage." The banter masked the tension of shared perils that had forged a bond between them.

"I need a slash!" Hulk excused himself, going back to his quarters to grab his last piece of kit. Returning, he found the slide-mount and board already fitted into the breaching module.

"Suit up, big fella," Kane ordered.

Though the breach suit took up valuable space, it was vital. Designed for both rescues and tactical insertions, the suit was a beast. Fully armoured, equipped with an array of pneumatics and hydraulics to assist movement and sealed against flame, smoke and chemical threats. Modular weapon mounts allowed for a variety of customised punches. Its two drawbacks are a thirsty power consumption and difficult manoeuvrability within the small areas of a spaceship.

The suit's bulk makes navigating confined spaces like corridors and doorways a challenge and even the simplest obstacles can bring it to a halt. However, the thicker the armour, the more punishment it can endure, giving the suit the appearance of the shell of a cannonball. But in Hulk's hands, the typically unwieldy armour moves with surprising agility, defying its cumbersome nature. The trade-off for the heavy armour, powerful weaponry and an array of electronic measures is energy. A mobile unit of this size can only hold a limited amount of internal power, which in a heated fight, soon runs out.

Hulk slid his communications cap onto his head before climbing into the suit, carefully aligning his limbs within their designated slots. Pope and Maihuni secured him in place by clamping braces and tightening the velcro straps.

"Hold up!" Hulk said, producing the team mascot. "Wouldn't want this little guy getting squashed!" Though it was foolish, no one could imagine their missions without Hulk's antics.

"Hand it over," Maihuni sighed, finding a spot in her medical bag for the small rabbit.

They then secured Hulk's shields, weapons and sensors. Maihuni tapped his visor, miming the closing of a lever. "Sealed!" she announced.

Hulk nodded, eyes scanning the room before issuing a calm but firm warning, "Powering on." Everyone instinctively took a step back, anticipating what was to come. With a mechanical hum, the suit shuddered and then whirred to life, its joints clicking and realigning with precise movements. The soft whir of small motors filled the air as they engaged, adjusting each limb slightly until it stood to a perfect attention. What had been an inert shell moments ago now stood tall and imposing, a towering figure of power, ready for action.

Pope stood before Hulk's visor, signalling the oxygen test. "Oxygen on."

"Sealed, oxygen levels full, green light." Hulk's HUD illuminated, revealing a multitude of system checks. He squinted towards Pope. "My vision is blurred though. All I can see is shit!" They both chuckled.

Maihuni released the slide board from within the module, which slid out and pivoted upright, lowering a foot plate. Pope gestured, mimicking the slide motion and Hulk slowly lifted his foot, initiating the suit's mechanics. He guided Hulk back onto the plate, who gave a thumbs-up and signalled standby mode. Temporary velcro straps were used to fasten Hulk to the slide board before Maihuni engaged the system, carefully lowering Hulk into a horizontal position. The board returned into the module and locked into place, leaving Hulk prone and secured for the trip to the Tyson.

Hulk occupied most of the space at the center of the module while Kane stacked supplies along the walls on either side. The rest of the team would have to squeeze into the narrow gaps between Hulk and the piled gear.

"Just lie there and do fuck all, why don't you?" Kane joked.

With Hulk and his breach suit dominating the module, space was tight, forcing them to cut their gear down to half of what they'd originally planned for the mission.

The pilot stomped on the slide-mount's locking lever, his body weight ensuring it was secured firmly to the floor. Hulk was going nowhere unexpectedly. Navigating through the stacked equipment, he slipped under the module's angular frame, avoiding dangling wires and slid into his seat at the controls, his sanctuary. He had made this move look easy, something mastered through countless repetitions. Hulk glanced up through his visor, amused at the thought of himself trying to be as nimble as the pilot, wondering if he'd even fit into such a cramped space.

The pilot shot a glance back at Kane, a look of eagerness in his eyes. He spun his finger in the air. "Load 'em up. We're close to moving out."

"Finish up, let's go!" Kane barked.

The team hustled, stuffing the last of the gear into the tender's module. It was always a delicate balance between what was important and what would fit. Brodie, Daniel and Davo squeezed in towards the back. With little space left but for the team, they mounted, stashing their rifles beside them. With one last movement, they shifted and repositioned themselves, searching for that elusive sweet spot of comfort.

Kane looked back at the crew in the Centaurus, observing their "better you than me" expressions as they watched them adapt to the cramped conditions.

The XO approached the hatch, his presence lending weight to Whiskey's mission. He offered a parting shot with an age-old Spartan quote, "Come back with your shield, or on it."

Everyone aboard knew they were merely the XO's instruments, sent to deal with dangerous, unwanted problems. "As if we're fools content to be no longer of this world," Kane thought, pride swelling as he gave a final thumbs up and the hatch sealed shut.

Every tender slated for the assault was now loaded and locked at their departure points, awaiting the command centre's signal. The pilot listened intently to the chatter in his headset, monitoring various channels and, most importantly, the ship's internal announcements. A buzzer wailed, followed by the chief engineer's voice over the coms: "All hands, this is the chief engineer. There's a maintenance issue in the mid-wheel and all drainage systems are temporarily closed. So, ladies and gentlemen, cross your legs."

"Cross your legs." That was the signal. The mission was on.

Pilots activated their flight plans. With radio silence initiated, communications ceased. They didn't know who or what might be listening and now was the time for quiet focus. Nothing needed to be said. Team Papa's tender launched first, followed closely by Team Quebec. Their timings were precise, synchronised, so they'd approach their breaching points together. Everyone adjusted their positions to get comfortable and began thinking about what was going to happen and what was expected of them.

"Hurry up and wait," Pope joked with a well-worn soldier's adage. He peered up into the tender's cupola but couldn't see much beyond the pilot's big boots. "Why haven't you installed the bar yet?"

With a thud Kane's fist struck the side of Pope's chest plate, perfectly placed where the armour jabbed into his skin. Just enough pain to make a point. Pope shot him a 'What the hell?' look.

Kane returned a stern, fatherly glare that instantly made him wonder what he had done wrong. He knew the procedure called for silence. Wide-eyed, Pope clasped his hand over his mouth, a silent admission of guilt.

The pilot, having completed his departure procedures, fidgeted nervously, checking and rechecking his gauges as his nerves took over. He pointed down to the locking lever. Kane, knowing that it was already secured, performed the required actions and gave a reassuring thumbs up.

With Kane distracted, Pope leaned over toward Maihuni and whispered out of Kane's earshot, "No matter how advanced we get, we always end up relying on our cave-dwelling instincts."

"Which caves? Earth or Proxima?" she shot back.

"Good point." He chuckled quietly, holding his breath to keep the sound contained.

The rush of preparation had settled into the familiar demand for patience.The old routine of 'hurry up and wait.' A flurry of movement and organisation, followed by the inevitable order to stand by. That unknown stretch of time, often longer than expected, yet suddenly cut short when least expected. Adrenaline faded into weariness and the boredom drifted to food, or a moment to read. With conversations forbidden, people usually dissolved into sleep. Only Pope struggled with the silence that was now enforced.

Uncertainty was normal for a soldier, but it didn't make it any less frustrating. With no room to move, no windows to see out of and a no talking policy in effect, the only thing to do was wait. Maihuni squirmed and pulled the small rabbit from her bag. The creature nuzzled her neck, catching everyone's attention. Davo offered it a bit of a protein bar, but the rabbit sniffed and rejected it, preferring instead to peer down at Hulk.

The other tenders were nearing or attaching to the Tyson. Once boarded, they would draw enemy teams in their direction and hastily establish their bridgehead, allowing Whiskey's approach to seem like logical reinforcements.

Once Six-Three-Charlie attaches to the Tyson, it would remain as a supply point for the team and, if necessary, an escape route. The pilot shifted to waiting, anticipating requests from the team. In the meantime, the pilots' time will drag, each second stretching out painfully as he waited for the results of the inevitable chaos. At that moment, Six-Three-Charlie remained connected to the Centaurus, patiently awaiting the signal from Papa One. The primary aim for Papa One is to bait the enemy. There would be a significant delay before initiating any signal.

Then it came, *"Stingray, this is Papa One. Requesting reinforcements, over."*

The XO responded, *"Copy out."*

*"Six-Three-Charlie, this is Stingray. Reinforce, over."*

With the wait finally ending, Trashie responded by detaching his tender, beginning its journey along the designated flight path. The systems hummed and directional thrusters intermittently engaged. Faces altered from a trance into a business-like heightened sense to razor-sharp focus. They had begun their approach to the behemoth intruder. Thoughts of family, pre-battle plans and what awaited them swam through everyone's minds.

"Thoughts, thoughts and more bloody thoughts!" Davo muttered, growing restless.

As the tender manoeuvred, the lights reflecting from Centaurus cut down through the cockpit. Battery cells clicked as they slotted into their rifles. They checked and rechecked the gear for the fourth or fifth time. Davo watched Kane and Pope, marvelling at how calm they seemed. Maihuni, unfazed, tucked the rabbit into an equipment bag hanging along the module's wall.

"Hurry up and wait, they say. How much bloody longer do we have to wait?," Daniel grumbled about the new term he had learnt. He watched Hulk with amazement, still snoring in his battle suit. Probably dreaming of his next hot meal.

The pilot guided them along their flight path, which eventually deviated toward the breaching point on the Tyson's hull. The increasing glow off the surface of the enemy liner marked their final approach.

"Come on, come on!" Brodie mumbled, not fully comprehending what lay ahead.

Kane was getting annoyed at everyone's failure to keep silent but said nothing till the red warning light flickered on. Shaking Hulk's suit until he woke.

"It's time." He mouthed.

Chapter 29

# POWER UP

The pilot nudged the tender's trajectory just enough to align it smoothly with the liner. Everyone fixed their gaze on the hatch, the gateway that could bring an end to their gruelling ordeal. A shallow clank reverberated as the two vessels met, sending an unexpected shiver through Maihuni. The tender settled and the breach began. Mechanical arms began their work with a chorus of metal scraping, followed by the groan of strained rubber seals. After a pause, they heard the sharp hiss of air as the outer seals inflated, expanding and gripping onto the enemy hull, like a parasite on a host. Kane wished the process would hurry. Every second increased the risk of the enemy catching on to their surprise assault.

The connection continued with a burst of sharp, metallic thuds as four harpoons punched through the liner's hull. Once embedded, the harpoons rotated into a locked position and pneumatic pumps whined, pulling the tender tighter against the liner. Valves along the harpoons released epoxy resin into compartments usually reserved for water and sewage. The epoxy reacted with the compartment's oxygen, solidifying into a rubbery seal, preventing any possibility of air leaks.

With the tender now firmly attached, the pilot called out, "Prepare for equalising!"

The team instinctively gripped their emergency oxygen kits, placing a finger from their free hand into an ear. The others quickly followed. Hulk, encased in his pressure suit, couldn't reach his ear even if he tried.

"Equalising!" the pilot shared.

Everyone took a deep breath, ears popping as they adjusted to the change. The tender's atmosphere synchronised with that of the liner, a critical step that could disrupt the mind if poorly managed. Kane noticed the pilot scanning the group, ensuring everyone was okay.

The ships were now unified. Four piercing clanks rang out, followed by a screech and the acrid scent of molten metals and resins.

"Outer ring secure!"

The breach unit drilled through the hull, its inner ring locking into place. Foam filled the space between the joined rings, solidifying and sealing the connection. A green light signalled the breach was complete. Only the secured hatch and tenders hatch stood between the teams and the liner's interior.

The circular hatches were wide enough for a fully geared operator or a slide-board bearing the might of Hulk in his breach suit. Although the interlocking doors were closed, a small central aperture allowed items to be passed through. A design feature meant for secure exchanges without fully opening the doors.

"Beethoven would have heard us by now," Kane said, suggesting even a deaf man could sense the commotion.

Kane clapped a hand on the shoulders of his team members, who instinctively formed a tight huddle over the top of Hulk. He motioned for the others to join as best they could.

"Our foes have shown how predictable they are. Now it's our turn to show them what a real team looks like," Kane rallied.

"If crossing this threshold is to be our end, we'll leave a lasting memory of the honour and bravery we shared."

"Honour and Bravery," the team echoed.

He met their eyes, one by one, sealing the moment with conviction. "Unleash the bugs!"

Now, with a surge of anticipation, the team readied for the next move. Their attention zeroed in on the hatch and the enemy awaiting them. Pope grabbed the first bug box and relayed its serial codes to the pilot, who paired the device with the tender. He then secured the box to the hatch's aperture at Hulk's feet, while Maihuni fastened it on the opposite side. They both gave a thumbs-up to signal that they had mounted and secured the box.

The box in question is the Centurion Drone Deployment System - Type B. Despite its unassuming grey plastic exterior, it housed a hundred synthetic nanobots, an arsenal of terror in any configuration. Whiskey rarely deployed drones, but when they did, Type B was their preferred choice.

Kane gave Hulk's helmet a playful smack. "You awake, numb nuts?" Hulk appreciated the moment of camaraderie. He then checked the slide-board for obstructions and gave Hulk the go-ahead. "Power up!"

Hulk lay flat on the slide-board, feet towards the hatch. As he powered up, his suit's joints gave a slight vibration, calibrating to account for the movements he'd made while it was offline. He was ready.

"Masks on," Kane ordered and everyone refitted their masks.

Kane did a final visual check and turned to Pope. "Okay, you ready?"

Pope nodded. "Ready!"

"Release the bugs!" Kane commanded.

Pope actuated the arming lever on the bug box and pressed the 'DEPLOY' button. He watched through the small window as the nanobots poured out. When the last of them had exited, he disarmed the lever. He and Maihuni released the latches and the box fell from the aperture, striking the heel of Hulk's suit. Sensors engaged, causing the mechanical knee to bend. Maihuni grabbed the box and tossed it aside as Pope reached for the backup. Together, they secured it to the aperture. "Box ready!" they called out and waited.

The team turned their attention to the monitor, eager to see the results.

The bugs denied the target space to the enemy. With minimal ballistic capabilities, they were ideal for breaching within a spaceship. The swarm operated cohesively, spreading out to identify and neutralise threats based on density and distance. Their strength lay in their collective ability to think and act as a unit.

The bugs infiltrated the Earthen ship like a relentless plague, buzzing and clapping as they fanned out in search of targets. As they ventured further, their effectiveness waned. Eventually, their tiny power sources drained, causing them to slow and drop from flight.

Twelve of the bugs acted as the operators' eyes, scanning the surroundings and relaying live footage. Three-dimensional imaging provided a layout of the area. The team watched holographic figures fleeing from the swarm. Although the furnishings were different, the recreational area mirrored the rooms on the Centaurus. The team noted other rooms that needed securing and their two planned exit points.

In a matter of seconds, several bugs soared higher, targeting the lighting with localised electromagnetic pulses (EMP). While not all bugs were successful, they neutralised enough light to trigger the relayed images being sent to thermal.

"Dance in the darkness, my friends," Maihuni said, her voice filled with anticipation.

To see the screen, the pilot brushed against Davo's gas mask, causing him to flinch. A handful of fleeing Earthlings made it to the exits, but the pursuing bugs caught some. As they closed in, their stingers flashed before they exploded. Each unloading a payload of gas or energised darts. A hit from one of those seventy-thousand-volt darts would leave a person incapacitated for a long period.

Visibility ranges, body counts, movement locations, temperature levels and contaminations data streamed in.

"Three down," Pope announced, relaying what the data already displayed.

The remaining bugs continued their search, buzzing around until they depleted their energy reserves and dropped to the floor. The images on the monitor flickered and vanished.

"Okay, Hulk. Pair your cameras to the tender," Kane instructed.

"Relayed," Hulk confirmed and the screen switched to Hulk's perspective, showing an image of Kane looking directly at the camera. They could now see what Hulk saw.

Kane felt relieved that a large welcoming committee wasn't waiting for them. The three motionless bodies outside were crew members of the ship. Although the enemy was currently focused on Papa and Quebec, they would soon turn their attention back to the threat to the ship's command.

"Let's get Hulk out," Kane yelled. He knew his voice was muffled by his mask.

Hulk flexed his feet and shoulders, easing circulation back into his limbs.

"Clear!" Pope confirmed, ensuring the space around Hulk's litter and the hatch was free of obstructions.

Kane began the countdown. "Three, two—"

"Wait!" Hulk interrupted.

"What now?" Kane looked down, exasperated.

"Just so you know, I'm only saving you from this burden, because I sense a whiff of snacks at their bar."

"For fuck's sake. Go!" Kane snapped.

With a firm pull, Pope released the latch. Both hatches swung outward as smoky fumes wafted in from the bugs' handiwork. The team shoved Hulk's slide-board through the hatch. It slid, extending horizontally to its maximum reach. With a click, the mechanisms fell into place, causing it to pivot upright as the side panels lowered.

Hulk was now free to create chaos. He stepped off and away from the footplate and the sideboard automatically retracted back into the tender. Pope sealed the hatch behind him.

Hulk raised his arms, his mounted weapons aimed toward the expected enemy approach. Sweeping his arcs, he fired three quick bursts of EMP, frying any electrical components in the vicinity. The remaining lights went dark. He paused, listened, nothing. No movement. No alarms. Focusing on his holographic overlays, his HUD displayed the three fallen enemies on the floor, still radiating heat. Activating his quantum stealth mirrors, he became almost invisible within the smoky haze, so long as he stayed still.

The team watched Hulk's cameras from the tender. Davo couldn't hold back his curiosity any longer, breaking the silence. "What's he doing?"

Kane leaned in, explaining, "He's setting the trap. You can't see it from our view, but his reflective plates make him mostly invisible."

Davo stared, clearly wanting more answers.

"We're expecting company," Kane said. "There's no point in us being out there with all the gas still around. Hulk has plenty of surprises for them when they arrive."

Moments later, Hulk's voice came over coms. "Movement." He remained perfectly still, hidden within the shadows. His

HUD lit up, showing five figures slowly navigating the rec room. Tension hung in the air as shadows and the threat of contact forced the enemy to advance with caution.

"Five armed. One is SERT." Hulk reported quietly, scanning again to confirm. "No masks and my readings say there's minimal contamination." The pilot's instruments confirmed it: 14.7 PSI, 78% nitrogen, 21% oxygen, with a bright green display reading 'AIR STABLE.' The ship's decontamination system had worked well. Kane gave the all-clear to remove their masks. A task they were all happy to follow.

Knowing that the enemy was getting closer, Hulk understood he could only unleash chaos for a short period before adopting a more conservative approach. He waited, hoping more would enter. The smoke in the room thinned and the enemy leader raised a hand, signalling the four others to halt. Two knelt as the leader lowered his goggles, scanning the room until his gaze settled on Hulk's location. Hulk's thumb traced tight circles on the trigger, eager to fire.

"Come on," Hulk muttered, craving the opportunity to unleash his fury.

The enemy's stare lingered and then he slowly raised his rifle toward Hulk's position. That was enough. With lethal intent, Hulk lifted his arms, the movement of the stealth mirrors briefly catching the enemy's attention. They froze, sensing their doom. Hulk didn't wait for a lock. He thumbed the trigger. A slight clicking of the weapon generated a scream within the moment of stillness. A blinding light flashed in sync with the crescendo, illuminating the entire room.

The first blast of his eight-barrelled arms roared to life, releasing a relentless rhythm of fire. The deafening whoop and thump of each shot filled the room as Hulk mowed down his initial target, then three more. Barely escaping the onslaught, the last of the enemy dove behind a counter.

Hulk kept firing, sparks flying as blasts ricocheted off the counter. "Enter now!" he shouted.

Maihuni moved first, pushing the hatch open and advancing and kneeling with her shield. Pope tossed out more shields as the rest of the team followed. One by one, they grabbed a shield and positioned themselves beside Hulk. The air was thick with residue from the damage caused by Hulk. Davo shut the hatch once they exited. This was a SERT only party. Hulk illuminated the counter with his spotlight.

"Stand with your hands up!" Kane shouted.

A tense pause followed before the enemy crewman slowly stood. Without warning, he raised his weapon. Kane barely managed, "Put—" before Hulk's triggers clicked.

The last blast sizzled through the air, barbecuing the man where he stood. His body crumpled to the floor. A hush of silence fell over the room as smoke curled in the glaring white spotlight. As Hulk's weapons cooled, they crackled, accompanied by the sound of bugs crushing underfoot. Kane stared at the body, not quite understanding why a member of the crew would choose death over surrender. A dangerous cocktail of adrenaline and bravado, he figured, but there was no time to dwell.

"What's the count, Hulk?" Kane asked.

"Five deceased, three stunned. No further movement."

"Brace those doors and help Hulk out," Kane ordered, pointing to the door leading to Tyson's command and the opposite that heads down to its spine.

Pope quickly braced the doors, while Maihuni helped Hulk. Hulk remained motionless even though his body screamed for movement after the long periods in the suit.

Kane banged on the hatch to let the others know to enter. "Brodie, Daniel, move the living into that room." They moved without hesitation, showing the value of the rehearsal.

"Centaurus, this is Whiskey One. Whiskey's in, light enemy over," Kane radioed. With their insertion complete, regular coms resumed.

"Sunray, copy, out." The XO's reply was swift, ordering Papa and Quebec to maintain pressure on the enemy, keeping them distracted from Whiskey's position.

Maihuni removed Hulk's armour plating and unclipped the suit's straps.

Kane asked, "What was left in the tank?"

"Nineteen percent," Hulk replied as he took his first step out.

"Shit. You could have put a few more blasts into that fella."

Hulk rubbed his eyes. His hair, matted with sweat that glistened beneath the suit's spotlight. He grabbed his kit from the tender and with a weary nod, fell in with the team at the door leading towards the ship's spine.

Frontline existence is mostly periods of waiting, broken by these sudden bursts of violence. Bursts where Hulk always proved his worth.

# Chapter 30

# SHOOT OR LEAVE

Having released the door brace that led to the ship's spine, Kane stepped forward into the dimly lit corridor, keeping his pistol steadily trained around the edge of his shield. Maihuni moved close behind, her eyes scanning over Kane's left shoulder, the butt of her rifle snug against her shoulder plate. They advanced single-file through the ship's narrow passageways, weaving through the clutter. Hulk and Davo followed behind. Davo, with a newly assigned role as pack mule, was lugging two unused shields and some door braces. His load made it hard to navigate some tighter spaces in the corridors. He cursed under his breath with each clumsy bump, wary of drawing unwanted attention from the Earthen crew. At the rear, Pope maintained a watchful guard. The fast pace made it hard for him to keep his shield raised, so he resorted to frequent glances, sacrificing security for speed. Clearing every room was out of the question as urgency dictated their pace. Locking down the ship's spine was critical to prevent the enemy from returning from the decoy provided by Papa and Quebec in the forward superstructure.

Some of Tyson's essential crew were still at their stations, but most remained in their quarters, having been trained well enough to stay put unless summoned. The team snaked through a series of left and right turns before reaching an emergency stairwell that will lead them closer to their goal of securing the spine's door. Kane stopped and quickly surveyed their surroundings, noting potential defendable points for their return, then signalled the descent to the lower level. Hulk took that moment to calm his furry companion, which he had retrieved from the tender.

Ahead of the team, two short corridors led to the emergency door that separated the superstructure from the ship's spine. Kane knew the enemy would send a team back down the spine to intercept him, so time was of the essence to seal the door before they arrive. Drills aboard the Centaurus allowed no margin for sloppy execution. Yet, the weight of their gear was taking its toll. The filtered light of the stairwell showed the heat that radiated from their bodies as sweat soaked through their suits. Davo's breathing increased in both speed and decibels. Already struggling with his fitness, he became less alert and the clanking of the shields echoed.

"We're almost there, big fella," Hulk encouraged, acknowledging Davo's efforts to keep up without complaining about his exhaustion.

As they exited the stairwell and pushed down the second-to-last corridor, two armed crew members suddenly burst around the corner, sprinting straight toward Kane.

"Contact!"

The team instantly dropped into a defensive stance. Kane knelt behind his shield, positioning his pistol, while Maihuni aiming her rifle around the shield's opposite edge. Hulk and Davo knelt behind Maihuni. Hulk swivelled and grabbed a shield from Davo. Pope, covering their six, faced his shield rearward. Their reaction synchronised instinctively. Kane's thumb brushed his safety switch, reassuring himself it was already off.

The two crew members stood dumbfounded, clearly unprepared for this encounter. The silence that followed was thick with tension. Fear flashed across their faces and they froze in place, surprised by the squad's poised aggression. They dared not raise their weapons, fully aware that Kane and Maihuni would swiftly eliminate them.

Kane, eyes still locked on his sights, barked a warning. "Shoot or leave!"

The crew members glanced at each other, then back at the heavily armed team crouched behind their shields. Their job descriptions certainly hadn't included armed conflict. Slowly, both lowered their weapons to the floor, took a few paces backwards, turned and ran.

Seizing the opportunity, Kane ordered, "Let's go! Move it!"

The team pushed forward, using the retreating crew members as a moving shield against any threats ahead. With a sense of relief, the door appeared. Maihuni and Kane took up defensive positions, covering Hulk as he forcefully swung the heavy-duty door shut, its ungreased hinges groaning from neglect.

Hulk unburdened Davo of a door brace, securing it across the frame. Davo dropped the rest of his load and set to work, altering the codes on the door's locking system. While the door wouldn't hold indefinitely, it would give them sufficient time to confront the ship's captain and coerce the enemy into backing down.

With the door securely locked behind them, the team retraced their steps back to the breaching point. There, they would regroup, rest and make final preparations for the impending assault on the command centre. Hulk unsheathed the large tomahawk from his waist, a formidable tool typically reserved for emergencies like fire and rescue. Today, however, it served a more deliberate purpose. Hulk swung the tomahawk as they retraced their steps, obliterating the electrical mechanisms of the corridor doors. Each strike sent sparks flying, sealing their path and ensuring the enemy will have many delaying obstacles.

The team moved swiftly, pausing only for the hiss of doors snapping shut behind them, followed by the heavy thuds of Hulk's tomahawk. The low hum of the ship's systems accompanied their march, underscored by the faint, almost eerie echo of music drifting from the crew quarters. A few curious crew members dared to poke their heads out, drawn by the noise of Hulk's relentless blows. A pointed rifle swiftly extinguished their curiosity, forcing them back into the shadows without a word.

Just short of the breaching point where their tender waited, Kane radioed Brodie. She removed the door brace, granting them entry. Hulk gave one last swing before leaning forward, placing his hands on his knees and catching his breath. Thankfully, they no longer needed to crouch in the dark to dodge unseen pipes. While they had been away, Daniel had repositioned the spotlight and battery cell from the drained breach suit, securing it to a more convenient pipe. As a precaution, Kane positioned the team strategically, covering both the door Hulk had just disabled and the one they would soon open to proceed to the command centre.

"Well done, everyone. Rest and resupply. We'll leave soon." Kane gave the pilot the all-clear to open up and began replacing their equipment. The team immediately appreciated the chance to rest, uncertain when the next opportunity would come. Everyone lowered their gear and attempted to relax, even if only for a short while.

The pilot handed out food packs and water. "What, no wine and cake, Trashie?" Pope teased the pilot.

"Only for those in first class," the pilot responded, barely mustering a smile. His long silent wait in the capsule had only heightened his nerves and his desire to be elsewhere.

"Feed the prisoners also, thanks, Trashie," Kane instructed.

Before distributing food to the prisoners, the pilot handed Brodie and Daniel their rations. They had already helped themselves to the food in the rec room, so they handed their rations to Hulk. He removed some lettuce for his mascot, sorted and placed the saved rations into the now emptied pocket. Kane didn't bother with interrogations. One of them was a SERT member, tight-lipped and trained to resist interrogation, making any information he provided unreliable at best. The other wore a chef's uniform and unless Hulk was also in the mood for a cooking lesson, the man likely had nothing of value to contribute.

They sorted their gear, scoffed down their rations and discreetly used corners as makeshift urinals. Kane set his rifle on the table and took a seat on a floor-mounted swivel chair, giving him a clear view of the rec room. Maihuni sat beside him, leaning in to check on the healing progress of his stab wound. The bodies from the breach still lay contorted where they had fallen, filling the air with the stench of death from released bowels. Kane suggested to Brodie to move the bodies to an adjoining storeroom. It was clear from her facial expression that she was not enthusiastic about the task.

Suddenly, Hulk's voice boomed with urgency, "Who's got some bog paper?"

Maihuni wrinkled her nose and raised her upper lip at what was about to happen.

"Here you go." Pope tossed him half a roll and a plastic bag. Hulk disappeared into the shadows of the storeroom before Brodie and Daniel began their horrid task. He took a quick look at the lifeless bodies as he passed, fully aware that he was responsible for their demise. Those soldiers did not differ from Whiskey. They had known duty, camaraderie and a fight they believed was just. Now, their bodies will be set aside, awaiting the moment when the living deem their removal a necessity.

Kane made use of the time to evaluate their mission's progress and was amazed to discover that his team had emerged unscathed. In his thoughts, he stared at the service area behind the counter, where a mural of earth adorned the wall. Over time, the painting had been chipped and scratched from stainless trolleys and plastic crates, leaving Oceania and Africa barely recognisable. Above the deteriorating artwork stuck a freshly made sign reminding staff to rinse glasses before placing them on trays.

"Why stick such an insignificant sign over a beautiful picture?" Kane mused aloud to Maihuni.

"Someone who wanted clean glasses, I guess," she replied, less invested in the moment.

Maihuni had noticed a change in Kane, his mind occasionally drifting, which was dangerous in their line of work. She could see his pain and wondered if the others noticed or if it was just her instincts that gave her insight into his state of mind.

She began gently, "You know, it hurts because she was willing to give her life to make things right with you."

Kane glanced briefly at Maihuni, then back to the mural. He stayed silent, absorbing her words. It was rare for those in their line of work to share personal thoughts, especially with subordinates. But Kane didn't tell her to back off and she took that as permission to continue.

"You won't forget her anytime soon and that's okay. She seemed like a hell of a woman." Maihuni then focused on the painting with Kane.

Kane reflected on Rayna and how his own vulnerability had become visible to others. He was surprised that nobody had made fun of it. Perhaps it was only Maihuni who noticed. Maybe they, too, cared for Rayna and were keeping their thoughts and emotions in check. Deciding to confront the unspoken, he spoke up.

"In the short time we had with Rayna, I let my guard down. I shouldn't have done that."

"It's normal," Maihuni reassured, both avoiding eye contact.

"She wanted to make amends," Kane reasoned, feeling complicit in her death.

"She did and that's why it hurts. We all make different decisions in the heat of the moment. She did what she felt was right. If she hadn't, she would have lived with the guilt."

Kane remained silent.

"Make the time you had with her count." Maihuni advised.

Kane turned to face Maihuni. In that moment, he didn't see her as just another soldier, but as a comrade sharing the burdens of their military family. He suspected the pain was familiar to her but chose not to pry. Her advice helped him acknowledge his mindset.

"Feels like I've been holding myself together with craft glue and duct tape and that shit ain't permanent." He spoke in a low voice, only for her to hear. "Thanks, Huni. That helped," he acknowledged.

"Then pull yourself together before the others catch on and start giving you shit."

"Yes, ma'am." He nodded, stood, and offered a hand to Maihuni, smiling as he helped her up. "I'm back."

Chapter 31

# CHAPEL LANDING

"*Sunray, this is Papa One. Sitrep, over.*"

"*This is Sunray. Send, over.*"

"*Multiple casualties. Three KIA, including Quebec One. Enemy strength is higher than expected. Requesting immediate evacuation, over.*"

"*Wait out,*" Sunray responded as he turned to consult with his team.

Whiskey heard the exchange over the radio and anticipated Kane's move. They stood as Kane raised an arm above his head, signalling his team to pack up and rally on him.

The XO worked closely with the tender pilots to ensure they were in position. He knew any mistake could spell disaster if the teams moved out of cover, only to find the tenders unprepared. Kane listened intently to the updates crackling over the coms, wondering if the XO was considering pulling Whiskey back as well. There was no such command. After a tense pause, Sunray finally responded. "*Papa One, this is Sunray. Withdraw to the Centaurus, over.*"

"*This is Papa One. Copy, out.*"

Once back on the Centaurus, Papa and the remnants of the teams would offload their wounded to the waiting medics, rearm and stand ready to be sent back into the fray.

**199**

No matter what they were currently doing, Whiskey knows it's time to get serious. Each member began their preparations. Packing their shit followed by kit checks, ensuring all equipment is functional and accounted for. Each member of the team checked their weapons, ensuring they properly loaded their power cells and engaged their safeties. Communication checks guaranteed their radios were operational and set to the correct frequencies. With door braces and shields slung comfortably over their shoulders, some took a hurried trip to pee, knowing that there might not be another opportunity for quite a while.

Making sure his vest and equipment fit properly, Kane moved to the porthole near the door that leads to the command centre. As he peered through, he anticipated seeing the breaching module to the left. Instead, his gaze was drawn to the shadowy horizon of the Jingai, faintly lit by the glow of a distant star. Kane squinted, closing one eye as the glare from the star faded as it slipped behind the asteroid. His attention shifted to the Bio-Stat, now drifting away from Jingai, its surface partially obscured but still highlighted by the star. In the distance, he could make out a tangle of mechanical snakes, their lengths poised, waiting for commands to enter the mine. Some repositioned slightly as their automated collision sensors detected the BioStat's approach.

"Not a sight I ever thought I'd see," Kane thought. Although he knew it had happened before, he couldn't recall a BioStat ever needing to detach from an asteroid.

He gestured toward the porthole, sharing the view. "Take a look." he said.

One by one, the team members peered through.

"It'll be back on Jingai soon enough," Hulk said confidently.

Pope, ever the analyst, added, "He's got a point. The company won't mess around when there's money on the line. They'll want that thing back on the rock ASAP."

Kane steered them back on course. 'That's only if we win this mess. So listen up.'

Fatigue was creeping in, dulling their reflexes, but they were still in it. Kane delivered a speech before stepping into the unknown—part command, part inspiration—the kind only a true leader could give.

"The boss picked us because we're the best. By the time we're done, they'll be writing training manuals on everything Whiskey achieved here." Now he was exaggerating and the team cracked smiles.

"Take note. If you're on the verge of dying, I'll drag you up and kick you in the balls," the team broke into laughter.

He then shot a look at Davo, "And for fuck's sake, make sure Sparky here doesn't make any of us look bad."

Hulk pivoted, grinning as he thumped Davo on his chest plate.

Kane turned to Hulk next. "And the only thing I want to see you hold is the enemy by their throat, not a fuckin' rabbit."

Hulk reluctantly handed his cherished bunny over to the pilot with a solemn warning, "It's your life," making sure Trashy knew to take care of his furry friend.

Kane continued, "So when this job is done, let's make it memorable and empty the Centaurus of all its mead."

"Oh, yeah!" Pope liked the thought.

"So, check each other's gear and let's get this done."

They quickly inspected each other's kit. Hulk ensured Davo carried the shields and braces comfortably. Each gave Kane a thumbs up. Pope released the brace on the door to the command centre, slinging it over his shoulder and securing the strap to his chest plate above his battery cells.

Hulk took the lead, shield in hand, wielding it as though it weighed nothing, his imposing stature deterring most would-be challengers. Kane followed, then Davo, Maihuni and finally Pope, once again, covering the rear.

*"Sunray, this is Whiskey One, over,"* Kane radioed in.

A brief pause, then, *"This is Sunray, go ahead, over."*

*"Whiskey One, chapel landing, over."*

*"Sunray, chapel landing, out."*

Capel landing, the code confirming that Whiskey was beginning the second phase of their mission. They moved at pace down the corridors toward the command centre, alert for any threats but not expecting any immediate resistance from the crew. If there were enemies to be met, they'd likely be SERT, defending the command centre.

# Chapter 32

# RIPPLED AND TWISTED

A doorway suddenly slid open beside Maihuni. Her instincts flared and she snapped into a firing stance, rifle up, finger brushing the trigger, ready to strike. Kane spun around; his weapon also trained on the emerging threat. The air crackled with tension. Not an enemy, but two wide-eyed teenagers stare back, their startled yelps echoing down the corridor. The closest, being a skinny, fair-haired boy whose name patch, read Jacob. Behind him stood his taller, older sister, Nalla. Without the height difference, they might have been mistaken for twins, both staring with a mix of fear and confusion. The teens were familiar with security forces, but not soldiers in unfamiliar uniforms pointing weapons at them.

Maihuni shot them a glare as cold and authoritative as a school principal catching truants in the bathroom. She was about to order them back inside when a radio transmission from Sunray crackled through her headset, requesting a situation report from Kane.

Sensing no threat, Kane turned away from the teens to respond. *"This is Whiskey One. En route short of chapel landing, over."*

Maihuni lowered her rifle, her ears tuned to Sunray's persistent commands. When she refocused on the teens, something felt off. Nalla was shifting, half-hidden behind her brother. A cold knot of dread tightened in Maihuni's gut as she spotted the glint of a pistol. Whatever the girl's intentions were, to protect her brother, or fight for Earth, Maihuni could sense the girl hadn't thought past this desperate moment.

Nalla raised the pistol and aimed but hesitated, trembling. When it came time to commit, she froze. Maihuni didn't. In a fluid motion, she brought her rifle to her hip and squeezed the trigger. Nothing. The rifle clicked and clicked again, the firing mechanism failing to connect with the power cell. Panic mirrored in both their eyes as their expected actions failed them. For a brief instant, Maihuni and Nalla shared a silent understanding that neither wanted to pull the trigger.

In a calm, motherly tone, Maihuni tried a different approach. "It's okay, we won't hurt you."

Tears welled in Nalla's eyes, but she didn't miss her second chance. Maihuni knew it, that exact moment when Nalla's expression shifted from sorrowful panic to cold, deadly resolve. The intense desperation in Maihuni's mind brought the scene to a painfully slow pace. Her voice broke free in a last-ditch plea, "Nooo!" She dropped her rifle and reached for her pistol. The hand grip instantaneously recognised her handprint, releasing it from her chest plate. She flicked off the safety, but Nalla's shot was already on its way. Knowing she couldn't draw her pistol fast enough, Maihuni fired her blast in desperation, striking nothing but the floor.

Adrenaline surged through her veins, but time seemed to drag painfully slow as she watched the brilliant flare of Nalla's burst approaching. The energy rippled and twisted, a chaotic blur that hurtled forward and crashed into her with brutal force. The impact hit like a sledgehammer, lifting her off her feet and slamming her back against the wall. She crumpled to the floor, every muscle failing her in the blast's aftermath.

The rest of the team turned, seeing her injured body coming to rest. They had no time to return fire, only to catch a glimpse of the door slamming shut in front of the boy's face. Jacob's screams continued through the door until his sister dragged him away. Hulk stayed at his shield while the others rushed to Maihuni's aid. Pope was quick to react and slammed a brace against the door, sealing it shut. He would deal with the little shits later.

Davo dropped the shields and pulled her slumped body into a less contorted position, away from the doorway. Kane spotted the trail of blood marking where she'd been dragged and quickly traced it to a wound at the base of her neck. He pressed hard with one hand, loosening her armour with the other to gain better access. Blood flowed through his fingers, pooling on the floor below.

"Sparky, put your hand here!" Davo slid his hand in, keeping the pressure as Kane tore away her field dressing and wrapped it around the wound.

Standard procedures were to use the wounded's field dressing, which proved to be useless. Kane unclipped Maihuni's med bag, which was stocked with far more supplies. Pope abandoned his position at the rear to assist Kane, quickly retrieving everything Kane might need from the med bag and laying the items out over Maihuni's chest.

Maihuni's eyes shot open, and in that split second, a surge of raw pain engulfed her, causing her body to instinctively recoil as she fought to push against it. Davo grabbed her arms, using his weight to hold her down and preventing her from thrashing. His blood-soaked hand slipped and he resorted to also using a knee. Kane administered a hemstat just below the wound. The drug quickly subdued her attempts to fight, leaving her lying there, dazed and trying to process her surroundings through a haze of medication.

"Steady, steady, don't move," Kane urged, trying to soothe her with a calming voice.

He tore open another field dressing, wrapping it tightly over the first and again pressed his hand firmly against the wound. The pressure was crucial to slow the blood flow and give the wound a chance to clot.

"Huni, I need to make this tight. Tap your hand if you can't breathe."

Overwhelmed by shock, the effects of the hemstat and Davo's weight, all she could manage were a few blinks to signal to Kane that she understood.

The second dressing was also a futile attempt at stemming the flow, which was spilling onto the floor once more. Maihuni, seeing Kane's blood-soaked hands, realised just how dire the situation was.

Pope's radio communications mirrored the urgency of the situation. *"Six-Three-Charlie this is Whiskey Three. To me, with a stretcher immediately, over!"*

The pilot rushed out of his hatch to the corridors. Sitting back on the rec room lounges, Brodie and Daniel's eyes followed his departure with a concerned look.

*"Sunray, this is Whiskey three, medivac over."*

The XO immediately responded to the medical evacuation request. *"Sunray, send over"*

Pope forgo the need for identifying himself, *"One serious. Request medivac, piggyback on Six-Three-Charlie, over."*

*"Acknowledged, out."* The medivac tenders were already in position just off the Tyson, ready to be dispatched.

Maihuni began mumbling softly, her words a mix of fragmented thoughts and speech. "Tell them it wasn't because I didn't try..."

Kane focused on preparing her for extraction and barely registered her murmurs.

She gazed at the vision of her parents above Kane and Davo, her voice faint but soothing. "Don't cry, I'm at peace." As consciousness slipped away, she tried to comfort the apparition of her parents.

Kane cleared the area where Maihuni was to be loaded onto the stretcher. He secured her arms and legs with bandages to facilitate the move and tied the used hemstat to the end of her bandage. A fresh hemstat was waiting for Trashie's arrival as he rounded the corner. Every second counted for her chance of survival.

They opened and braced the stretcher. "Lift," Kane instructed as they centred her limp body and secured it with Velcro straps.

Kane pointed out the hemstat to the pilot. "If there's any delay and if she acts up, use it."

"I've got this," the pilot replied with a reassuring nod.

Davo's bloodied hands grasped the stretcher's handles. As they lifted, Maihuni searched for Kane and whispered, "I'm sorry."

In that fleeting moment, Kane couldn't find the words to say how much he'd miss her. All he could manage was, "Don't be. Just come back."

They watched as she swayed back and forth as they ushered her out of view.

"Your thoughts?" Hulk asked.

"I don't know," Pope admitted, his gaze heavy. "The odds aren't in her favour, but I've seen people survive worse.

I just hope this isn't the last time we see her." Kane said, his anger clear. "But I hope I'm wrong."

He collected the medical supplies and packed them back into Maihuni's bag. The items inside the bag became soaked in blood, which had pooled at the bottom. He wiped his hands on his trousers, then grabbed his dagger and used the point to pierce a hole in the bag. Blood oozed out, splattering into the pool on the floor. Kane questioned himself. Was there anything more he could have done? With the resources he had, there was little else he could have. Her life was now in the hands of those with greater skills than his.

They cleaned up and prepared to move on. The enemy would by now be breaching the doors from the ship's spine, so time wasn't to be wasted.

"What about the two shithead kids?"

"We'll come back for them," Kane said firmly.

They all understood that the mission took precedence over personal vendettas. His aching heart wouldn't win wars, so he stood tall and carried himself with the professionalism his team expected.

Continuing to the command centre, Kane's mind raced. He remembered Maihuni's warm, maternal nature and wondered why someone with such a passion for healing would choose a profession that dealt in pain rather than one that cures. She should have been saving lives, not caught in a gunfight. Maybe she had her own demons. After all, we all do.

He recalled Maihuni's earlier advice about Rayna. "We all react differently when making decisions. The actions she took were because it was the right thing to do. If she hadn't, she would have lived with the guilt." Like Rayna, Maihuni was where she was meant to be and he knew his time with both had been an honour.

Ultimately, the team knew they were just numbers. The mission had to be completed, but now they would face the challenges with one less member.

Chapter 33

# A BRETHREN

"It always has to be difficult," Hulk thought the moment he heard the noise.

"Boss, we've got company." The rattle and clank of heavy footsteps echoed from far ahead.

Kane snapped into action, listening intently, evaluating the noise and sizing up the situation. "It sounds like there are many voices gathering. I'm guessing they're outside the command centre," he said to Hulk.

"SERT wouldn't be that sloppy. They're most likely crew," Hulk added.

Whiskey was under-manned and Kane had had enough of playing nice with these people, especially after the shooting of Maihuni. He was ready to give their uninvited guests a surprise they wouldn't forget.

He withdrew Hulk and Pope down the corridor, positioning them in a prone position.

"Prepare for an ambush, front," he instructed, pointing down the corridor. "And move against the walls so there's space for me."

Hulk and Pope exchanged a quick, puzzled glance, not entirely sure about Kane's choice of defensive location, but they followed orders. They pressed themselves against the walls, squeezing as much as possible behind the piping that ran from floor to ceiling. They lined up spare battery cells for easy access and settled in as comfortably as their body armour and the grated floor allowed.

Kane grabbed Hulk's shield and sprinted toward the next bend in the corridor.

"What the fuck is he doing?" Hulk whispered, watching him with total confusion.

As he retreated down the corridor, Kane swung the shield, obliterating the lights in his path. He continued past Hulk and Pope until he finally reached the far end behind them. He then placed the shield in the corner, redirecting the light from around the bend onto the ceiling in front of Hulk and Pope, masking their position in a deep shadow.

Satisfied with his handiwork, Kane rushed back, passing the two on the floor to position himself at the corner where the enemy would approach. The growing sound of boots told him they were getting close. He quickly turned to check the scene from the enemy's point of view. All they'd see was darkness, daunting, impenetrable darkness. The shadow beneath the light reflected off the shield kept Hulk and Pope hidden, invisible to the enemy's eyes. Pleased with the setup, Kane returned and took a prone position as the darkness swallowed them whole.

As Kane settled, he whispered into the radio, instructing the tender pilot to delay Davo's return until he gave the signal.

Then, with a calm demeanour, he gave orders. "Don't move. Don't talk. Stay low and wait until I fire." The three of them lay in silent anticipation.

Pope couldn't help but marvel at how the boss' mind worked under pressure, always being sharp and decisive when immediate action was required.

The echo of footsteps grew louder, their shadows stretching around the bend in the corridor ahead. The team made minor adjustments to prepare themselves, raising their weapons to aim. Kane squinted with one eye, lining up his shot, knowing he would fire first. Elongated shadows shortened as the figures neared, their legs casting a strobing, ghostly dance along the walls. Kane counted them, one, three, six… then stopped counting as the numbers swelled to possibly ten. They charged forward, only slowing as they turned the corner into the darkened corridor.

"Come on, a little further," Kane willed them silently.

The team remained motionless. That nerve-wracking moment lingered, where everything could go south if the enemy spotted the ambush too soon. The lead figure slowed, prompting the others to follow suit. He must have sensed something was off, his gaze fixed on the light deflected by the shield. As Kane watched their silhouettes, every calculated move became apparent. It assuring him that the deception was working as planned. Their sprint had turned into a cautious walk as they struggled to comprehend the crunch of the lights glass beneath their boots.

Pope felt a bead of sweat slide down his forehead, pooling at the edge of his brow. His thoughts raced, cursing Kane for holding off. "Come on, come on, fuck ya'. Pull the bloody trigger," he thought, frustrated by the delay. Even though he knew Kane was waiting for the perfect moment. That moment when the enemy were in a position to achieve maximum carnage.

The lead figure hesitated, slowly raising his weapon. That was the last move he ever made. The reason they were sent rushing towards their enemy didn't matter. They were just in the wrong place at the wrong time, destined to become Whiskey's latest casualties.

From within the shadows of the corridor, three streams of laser fire erupted, cutting through the crew. Bodies twisted and fell as weapons clattered to the ground. Whiskey reloaded and fired again into the now prone figures. Flashes of successful strikes illuminated the bouncing carcasses thudding against the grated floor. They never stood a chance. The fierce sounds of the firefight gradually faded, replaced by the familiar hum of running sewer pipes. No thoughts lingered on the dead. Instead, the three changed cells and exchanged adrenaline-fueled smiles, the kind only shared within the brotherhood of combat. A sense of purpose forged in the face of danger. Yet, the victory only added to the growing tally of unnecessary casualties.

They waited, scanning for movement, but there was none, nor did they expect any. Kane wondered why someone had unnecessarily sent them to their deaths. They would have been far better off defending their command centre.

"Oh shit," Kane muttered as footsteps echoed from behind.

The three rolled onto their backs and aiming their rifles toward the direction of the shield. From around the corner, Davo appeared, returning from the tender. He froze at the sight of three guns trained on him.

"You're one lucky fucker, Sparky. You were nearly toast!" Pope said, lowering his weapon.

The message to wait must have got lost in the fog of war. Kane didn't dwell on it.

Narrow-minded, self-serving individuals who used friendships for their own gain had surrounded Davo's life. A life he had accepted until he met Kane. Now he knew the type of people he wanted to be around. It wasn't the danger, the blood, or the death that excited him. Those filled him with fear. It was the bond of camaraderie that truly stirred him. They were like family. A family that would take you on that scary roller coaster or stand between you and a ravenous dog. Brethren that respected you with no strings attached. And for Davo, being a part of this family was more than enough.

When the team moved back into the light to regroup, Kane noticed that Davo's clothes were soaked in Maihuni's blood. The deep crimson soaked through, causing the fabric to cling tightly to his skin, almost fusing onto him and creating a grotesque, sticky sheen that reflected the dim light. Kane knew this would become yet another haunting memory, one that would wake him from sleep on countless restless nights.

"Did Maihuni make it?" He asked.

"Sorry, I don't know. Medivac took her as soon as we got to the tender." Davo replied, still trying to wipe Maihuni's blood from his hands.

Kane accepted the response, knowing Davo had done everything he could. For a civilian, Davo had more than earned Whiskey's respect.

"Thoughts?" Kane asked everyone.

Pope reiterated, "Why would he make such a stupid move, sending those guys to their deaths?"

Kane could only agree as he stowed his depleted cells.

"Okay, let's keep going."

Hulk collected his shield and they advanced, stepping cautiously through the dark corridor littered with the enemy's fallen. The acrid scent of charred flesh and exposed innards hung in the air. Hulk covered his mouth and nose with his hand, navigating the maze of contorted bodies and making sure not to slip in the pooling blood.

"They're all crew," Pope observed.

"This one's the ship's XO," Kane added, pointing to one of the fallen.

Navigating past the mayhem, they reached the end of the damaged lights and paused, taking a moment to breathe in a better version of recycled air. Hulk unclipped his water bottle and washed the acidic taste from his mouth. Then asked, "Why didn't they send SERT?"

"Hopefully, there are none there," Pope speculated.

"I'm betting you're right," Kane agreed. "That's why our other two teams struggled. Quebec and Papa's diversion must have pulled all the enemy teams toward the forward superstructure."

"But that still doesn't explain why the XO came after us instead of defending the command centre," Pope remarked, still puzzled.

Kane motioned for Hulk to keep moving. "Let's go ask the captain that very question."

Chapter 34

# TALK TO THE BOSS

"*Whiskey One this is Six-Three-Charlie, over,*" the tender pilot radioed Kane.

Trashie didn't wait and with an urgency in his voice repeated, "*Whiskey One, this is Six-Three-Charlie, over.* "

"*Send?*" Kane waited.

Kane patted his hand down, signalling everyone to take a knee. They spread out and Pope moved back to the last corner to keep a watchful eye on their rear.

"*We have guests knocking on our door, over.*"

"*Wait out!*" Kane took a moment to pre-empt his chess moves.

Hulk and Pope watched their fronts, waiting for Kane's instructions. He weighed their options. If Whiskey returned to the breaching point to defend against the enemy coming back from the forward superstructure, they would likely be overrun or forced to withdraw via the tender. Either outcome would mean mission failure and give the enemy the upper hand. If they pushed forward to the command centre, there was still a chance of success without resorting to a last stand firefight. Their best hope lay in capturing the command centre and, ideally, the captain. It was the only option that could avoid complete failure.

Kane gathered his team. "Alright, we stick to the mission. Maihuni getting shot won't be for nothing."

He radioed Trashie. *"Secure the prisoners, detach the tender, and release oxygen into the breached room."*

There was a pause before Trashie's reply crackled through. *"Can't detach unless we seal both the internal and external doors, which means I can't release the oxygen. Over."*

"Fuckin' safety features," Kane muttered under his breath. Regaining his composure, he responded, *"This is Whiskey One, copy that. Leave the prisoners and commence extraction. Out."*

"Let's move!" Kane ordered. They reached the final corridors leading to the command centre entrance.

Brodie and Daniel left the prisoners in their secured room. They then boarded the tender, which separated smoothly from the liner, its thrusters flaring briefly till the tender withdrew to a safe distance, hovering just beyond the reach of the ship's shadow. The pilot could see the immense mass of the liner, its exterior marred from the addition of their breaching door. They monitored the radio, ready to respond the moment new orders came through.

Kane heard the transmissions between Trashie and the Centaurus and knew the tender was safe.

"Sparky, it's all yours," Kane said, gesturing to the door's locking system.

Davo immediately went to work, removing the faceplate. His fingers twisted and squeezed wires with practiced precision. He quickly hard-wired it to his workstation.

Kane paused for a moment to collect his thoughts before making the entry. He closed his eyes briefly, picturing the layout from their rehearsals on the Centaurus and mentally bracing himself for what lay ahead.

"When we open the door, Hulk will lead the way with his shield facing forward. Pope and I will handle any threats in front. Davo will cover our rear. Once we've dealt with any immediate threats, Hulk will advance while we clear the flanks."

Kane issued his last command before they moved in. "Remember, we're not here for the captain or his crew. We're here for the ship. If anyone resists, neutralise them!"

Stepping aside, Davo then announced. "Got it! Just press the open button."

Hulk positioned his shield in front of the door. With his rifle aimed over Hulk's shoulder and the shield, Kane reached out and hit the open button. He pressed once, then twice and again. Nothing. The door's mechanisms whined as it fought against the obstruction, vibrating the frame slightly, but it remained closed.

"Door braces?" Pope suggested, suspecting the enemy had barricaded it from the inside.

"Seems like it. Alright, Plan B," Kane said, though he hadn't yet figured out what Plan B would be. After a moment of thinking, he turned to Davo. "Hey Sparky, what's the quickest way we can talk to the people inside?"

Davo mulled over the problem and suggested, "Maybe try the emergency coms down the corridor?"

Kane hadn't considered that option, but it made sense. He hustled down the hall until he spotted the yellow and red markings on the wall. Beside an emergency axe was the coms panel. The operation was straightforward. He flipped open the cover and pressed the button, knowing someone in the command centre would follow protocol and connect.

The intercom crackled to life. "Command. State your emergency."

Kane spoke calmly, "This is Sergeant Kane from the Centaurus. We're taking command of this ship. You are to open the door. Consider your crew's welfare as the emergency."

There was a brief pause before a reply, laced with defiance, came through. "The captain wants you to know that this ship already has a commander. Kindly refrain from using the emergency coms unless there's an actual emergency."

"Cheeky prick," Kane muttered under his breath.

He leaned in and replied, "Please inform your captain that we'll use explosives to open that door. Save your crew from harm and open it."

There was a longer pause as those in the command centre weighed their options. Finally, the intercom crackled to life with a female voice. "Sergeant Kane, this is Captain Ainsworth. As you know, it's my duty to defend this ship. We will not be surrendering her to you."

Kane felt momentarily puzzled as he noticed that the captain's name differed from the one given in the briefing. Not that it mattered much. The intel was probably outdated. He weighed his response carefully. The idea of the command team trying to defend the ship was suicidal. Maybe the captain was stalling, hoping her remaining security teams would return from the forward superstructure in time. They were definitely en route.

Then it clicked. "It's those teens!" he muttered.

Kane returned to his team by the command centre door for their opinion. He recalled the incident when Maihuni had confronted two teenagers in a doorway and couldn't remember their first names exactly, but he was certain their surname on their suits was Ainsworth.

"They were siblings," Kane speculated. The team listened intently, trying to piece together where he was heading with this.

"And guess who's in command of the Tyson?" Kane prompted.

Hulk was quick to connect the dots. "They're the captain's kids?"

"He wouldn't have been happy hearing that we have them." Pope smiled.

"She doesn't know yet," Kane corrected.

"So, are we getting mummy to open this bloody door, or what?" Hulk snapped, growing frustrated with the slowed progress.

Kane considered the sequence of events that he believed had unfolded after Maihuni was shot. He assumed that the two teens had retreated into their room and reached out to the captain. Naturally, as a mother, the captain took immediate action to protect her children.

"That's why she sent those crew members," Hulk concluded, understanding the situation.

"What kind of captain puts her crew in harm's way like that?" Pope grumbled, clearly unimpressed by her actions.

Kane continued, "She wouldn't have known about the ambush." The XO likely thought they could stall us until their security forces arrived. He paused, lost in thought. "Who knows?"

The team quickly discussed adjustments to their plan, considering this new information. Kane had Hulk and Pope return to the teens' doorway and stand by. They retraced their steps into the dark corridor and over the motionless bodies. As they reached the light on the far side, they exhaled deeply, relieved to take in the fresher air. A large puddle of blood surrounded by the prints of the earlier chaos and a brace marked which door the teens were behind.

Kane stopped at the emergency coms, hoping the captain might reconsider his terms. He prepared to contact her again, noticing his own bloodied fingerprint on the button. He glanced down, for the first time noticing his blood-soaked uniform and deciding to use his fist to press the button.

The intercom crackled to life. "Command, state your emergency?"

With Maihuni's loss weighing heavily on him, Kane cut straight to the point. "Get your captain on the line, you fucking idiot!"

The operator didn't engage and soon after, the captain's voice came through.

"Can I assist you, Sergeant Kane?" The captain's tone was calm, but Kane could sense the distress it masked.

"Yes, you can," Kane replied sharply, then laid out his demands. "I have a team at your children's door." He paused, listening for a response, but her silence only confirmed his suspicion about her relationship with the teens.

"As you're probably aware, they shot one of my soldiers." He paused again. Still, there was silence.

"Since your children are armed, my team has two options." We can either breach the door and neutralise the threat permanently, then coming back here and do the same in your command centre. Or..." He paused for emphasis. "...your children can open the door and surrender. If they surrender, I'll bring them to the command centre, where you will also surrender. This second option ensures no further loss of life among your crew."

Captain Ainsworth paused to discuss their options with her team. Despite the grim choices ahead, Kane believed she wouldn't sacrifice her children. He also hoped the security teams wouldn't arrive in time to save the ship.

Her decision came moments later.

Kane joined the others at the teens' door. His eyes glanced at the puddle of blood reflecting the corridor's lights. The presence of field dressing wrappers and a chaotic mixture of bloodied hand and boot prints evidence of the frantic attempts to save Maihuni.

Behind the door, the captain's voice crackled through the intercom, speaking to her children. Her calm words soothed them and the girl finally lowered the pistol, letting it fall onto her bed. The siblings held hands as they opened the unbraced door.

"Turn around. No struggling. No talking," Kane warned. "Follow our instructions and you won't be harmed," Pope said as he cuffed them.

The teens' shoulders slumped in reluctant acceptance, their eyes darting nervously.

Suddenly, a burst of static erupted in Kane's earpiece, followed by a tense, urgent transmission.

*"Whiskey One, this is Six-Three-Charlie. Sensors have detected movement in the breach point, over."*

*"Copy, out,"* Kane responded, then refocused on the teens. "We're taking you to your mother." Their nods were a mix of fear and guarded hope.

He knew the enemy security teams were a genuine threat, so they had to move fast. "Let's go!" Kane's voice was strained with urgency and the team retraced their bloody footprints down the corridor. Once again, the corridor's lighting dimmed, replaced by darkness. The stench of charred flesh grew unbearable, thickening the air and making the teens gag. Their steps splashed through pools of blood as they advanced. Their flickering torchlight cast eerie shadows on the faces frozen in death. Each face that passed in the torch's glow was a haunting reminder of the lives lost.

Nalla collapsed, her body crumpling towards the blood-soaked floor. Pope's reflexes kicked in, catching her before she landed in the sticky pool of crimson. "Sparky!" he shouted, his voice tinged with urgency as he tried to minimise his exposure to the toxic stench.

Davo, with his bulk, was the obvious choice for carrying Nalla and he immediately understood his role. He swept Nalla up, throwing her over his shoulder. Securing her by locking her knee and wrist in his grip. He straightened and gave a decisive nod and they continued.

Pope turned to her crying brother, gripping his thin upper arm with a firm grasp. "This is no time to crack the sads. Stop crying and focus!" he snapped, shoving him forward with a sharp push to keep him moving. His eyes flicked back constantly, knowing the Earthen SERT would clear the doors fast.

"Hurry the fuck up!" he hissed urgently

Reaching the shadows, Kane guided Davo and the boy through the tangled bodies.

"Contact rear!" Pope shouted, his voice cutting through the tension.

Caught in an unwinnable position, everyone briefly spiralled into a panic before professionalism took over. Their training kicking in instinctively. Pope dropped to one knee and drew his weapon, locking eyes with the enemy's lead scout who mirrored his move. They took aim at each other, a tense standoff locking them in place. Neither fired. Pope didn't want to ignite a firefight, and he suspected the enemy was just as hesitant. Maybe they weren't sure who he was.

With no other option, Hulk planted himself defensively amidst the fallen, while Davo hurried the two teens toward the next corner. Kane and Pope then took precarious cover behind Hulk and his shield, preparing for whatever came next.

Opposing Whiskey, the enemy moved with precision, advancing behind their own shields to form a protective barrier around their scout. It was clear to everyone Whiskey wasn't getting out alive. They accepted their fate without hesitation. If they were to join the fallen, they'd do so on their terms, guns blazing. They would find a grim satisfaction in taking down as many enemies as possible.

"If our numbers are up, I'm leaving with an explosive farewell before I'm released into the dark void." Pope shared.

Kane whispered to his comrades, "They know we're finished. Why aren't they firing?"

Pope replied, "Maybe they've caught a whiff of Whiskey and are now scrambling to figure out how to dodge their own demise."

The answer came from behind. Earlier skirmishes had disrupted communications between captain Ainsworth and the Tyson teams. From their last exchange, she knew they were heading back to the command centre. With coms now reestablished, she realised her teams were closer than expected. Her children would soon enter the corridors with the enemy, so she rushed off to make sure they wouldn't get caught in a crossfire. Before leaving the command centre, she instructed her teams to not fire upon the enemy unless given explicit authorisation.

She ran out of the command centre and had reached Davo, who was holding her cuffed children.

As soon as she saw her mother, a whimpering Nalla cried out, "Mum!"

Davo fumbled for his pistol, ready to confront the unexpected figure sprinting down the corridor. The captain slowed and raising her hands, offering a token of surrender.

"Please, I'm the captain," she said, identifying herself. "My ship is yours. Please, just release my children."

The offer of an Earthen ship was unexpected but Davo knew it would be short-lived. Kane's wrath would be worse and he knew that keeping the teens was their only leverage in the unfolding chaos. He refused to let them go without Kane's approval.

"You'll have to talk to the boss, but he's busy," he said, nodding his head to show the direction.

The captain didn't hesitate and edged her way past and continued towards the corner. "I wouldn't," Davo warned. "Your team just showed up and a shitstorm is about to kick off."

She had surrendered her ship for the sake of her children and intended to keep her word. As she rounded the corner, a grisly scene greeted her. Bodies lay scattered across the floor, shrouded in shadows that hid the blood streaming down the corridor. A wave of nausea hit her as her heart sank. She knew she was the reason for these deaths. As she passed, her eyes scanned the motionless bodies, searching for any signs of life. There was no movement among the fallen. She spotted Kane and his men positioned behind their shields. In the dim light beyond, her teams lined up, forming a solid wall at the far end. The atmosphere crackled with the anticipation of conflict, every figure frozen in place, waiting for the inevitable clash.

To Kane's surprise, a tall, well-dressed woman stood next to him and the wrinkles on her forehead revealed the stress she carried despite her age. Her pale face was framed by a short bob, streaked with grey that she had made efforts to conceal. Her skin had a hardened texture, like stale bread and her thin, almost translucent arms revealed veins that traced the contours of her aging bones. Despite her frail appearance, her uniform was immaculate, smartly pressed and perfectly tailored. Indicating that while her body was faltering, her mind remained sharp and unwavering.

At first, it seemed unbelievable that someone would walk upright into a soon-to-be gunfight. Then it hit Kane. She was most likely the captain. Whiskey heads turned, eyes tracked her every step as she edged past their shield.

"Sargeant Kane." She greeted.

"Captain." He responded calmly, as though they weren't adversaries.

Stopping in front of Hulk's shield, she faced Whiskey's enemy. Her men.

"I'm captain Ainsworth." She identified herself, knowing she was in darkness.

Kane couldn't believe the luck that stood in front of him.

"Jesus Christ, skipper, what are you doing?" was the response from down the corridor.

The captain recognised the voice. "Corporal Barnes, I've surrendered the ship. You must lay down your weapons to these men."

Silence followed and in that brief pause, countless scenarios raced through Kane's mind. The worst included their captain going down with them.

"Just a few inches to the right." He thought. If the worst happened, he wanted to get one good shot off before his end.

Finally, the response came.

"We'll wait for the sergeant, Skipper." Corporal Barnes understood this was a fight he could win, but he also knew it wasn't his call to make, nor the captains.

The captain turned and glanced down at Kane. "We'd better go, sergeant." She moved back behind Hulk's shield, stepping cautiously.

She covered her mouth and nose, refusing to look at her fallen crew again. Once she was clear of the dim light, she paused and turned to look at the bloodied footprints trailing behind her. She imagined them not as her own, but as the marks left by those she had ordered to protect her children.

Whiskey followed slowly, stepping backwards, with Hulk's shield providing a mobile barrier against the weapons aimed at them, until they turned the corner. The captain and her children clung to each other, reluctant to let go. A firm shove broke the moment, shifting their focus. They moved toward the command centre, where Pope sealed the door from the inside.

Chapter 35

# MY CONDOLENCES

As the team entered the command centre, they moved through a narrow hallway. A door on the right led to the captain's briefing room, while the ones on the left opened to the captain's and XO's quarters. Pope stood guard as the captain guided her children safely into her quarters.

The command centre itself was at the end of the hall. This is where the captain kept track of the ship's status and delegates duties. The arrangement is unchanged when working from the other superstructures, only the crew are different. Every sound and vibration in the command centre are potential indicators of the ship's condition. The room is designed for monitoring and control, with four operators at their stations. Their training and professionalism was evident as they barely acknowledged Kane's entrance. One briefly glanced over, but seeing Kane's blood-soaked arms, he refocused on his gauges. There was a sense of tension and urgency in the air as they worked with stoic precision, carefully monitoring the panels and walls filled with switches and knobs. The steady ticking of quiet alerts, sporadic status indicators and composed transmissions conveying orders filled the room with a sense of importance.

The room had a strictly practical design, keeping the lighting low to minimise glare on the monitors. There was a swivel chair at the rear wall for the captain, allowing her to oversee all activities. Behind her chair, storage compartments were stocked with manuals and schematics, all clearly labelled for rapid access.

Given the circumstances on Tyson, communications were restricted to essential information, with only key crew members on duty. Hulk, Davo and Kane gathered in the briefing room to discuss their next move. Their orders from Centaurus were clear: secure the captain and continue defending the command centre. However, Centaurus' radio communications had yet to mention reinforcements, something the team had been hoping for.

Pope entered briefly, drawn by the sound of the captain's tearful conversation with her children. "Here's the heads up," he whispered. "The XO lying face down in the corridor is the teens' father."

"Shit," Hulk exclaimed.

Kane turned to Pope. "Keep an eye on them. The girl has already shot one of us and I wouldn't blame her if she did it again. If the captain leaves the room, lock the kids in."

"Got it," Pope replied.

The three took a moment to appreciate the unexpected comfort of sitting in the briefing room, nestled into the cushioned chairs, taking advantage of the snacks on the table.

"Why is it the higher up you are, the better the food you get?" Hulk asked.

"People like to please the captain, I guess," Kane responded.

"More likely, the captain has refined tastes and no one would want to feel her wrath." Pope had a dig at what he perceived as aristocracy.

Kane played along. "Possibly that too."

Davo added, "You know I should be in that chair." He pointed to Kane, who found comfort at the head of the table.

Kane took the bait. "Why's that Sparky?"

"While you three were playing wars, the captain handed over the Tyson to me." He smiled.

Calmly, Kane gave a response, "While we were out risking our necks for Proxima's honour, you, smartarse, were hiding around the corner playing hide-and-seek with your teenage entourage."

Davo relented, saying, "Fair call." The smiles on the faces of those in the room reflected their delight at how well he was integrating.

Kane began laying out the situation as he saw it. "If the enemy doesn't comply with the captain's orders, we need to be ready."

Davo frowned. "Surely they won't attack the command centre with their own people inside?"

Hulk shook his head. "We wouldn't. Not if it meant hurting our own."

"We haven't seen the Schmeller yet. If he is in charge, who knows what he'll do?" Davo questioned.

"Good point," Kane nodded, considering the implications.

From the doorway came, "Shall I join you?" the captain asked.

Kane was surprised she hadn't spent more time comforting her family, yet she was a professional who took her duties seriously. An open hand gestured for her to join. Meanwhile, Pope secured the captain's quarters and joined the team.

"My condolences," Kane said as he stepped away from the captain's seat. She nodded in silence as she took her place.

"You're still the captain of this ship and you need to run it as such," Kane said firmly as he took another seat. "To avoid disruptions, your crew needs to know you're still in charge, but they should remain in their quarters until further notice. Although any decisions regarding ship movements or security must go through me first."

"Thank you, sergeant. I'll do what I can to ensure a smooth transition," she agreed.

Kane nodded and pressed on. "My duty is to protect the command centre and everyone in it, so we need to address the situation with your security team."

"As you know, I've ordered them to stand down." She clarified.

"And?" Kane prompted, wanting more details.

"You understand, there's a fine line between commanding the ship and overseeing security operations. That was the XO's role. And, as you know, my husband is no longer with us."

Kane had already expressed his condolences and he wasn't about to revisit the issue. He also sensed she didn't expect him to.

"So, what do you think your security teams will do?" he asked.

"As you saw in the corridor, they were hesitant to follow my orders on this matter. They respect me, but they fear warrant officer Hill, their commander."

"The Schmeller," Pope quipped.

With a smirk at the reference to Hill's unmistakably overbearing nose, the captain nodded. "Yes, I believe we're talking about the same person. My husband kept him in check, but Hill can be quite the unsympathetic egocentric."

Kane leaned forward. "What do you think his next step will be?"

"He'll want to position himself as the saviour, not as the man who surrendered. I believe he'll try to retake the ship. His men will follow him, mostly out of fear."

This aligned with what Kane had expected, so he began planning his response accordingly.

"All right, captain, I need you to broadcast this message to the entire ship," he said, scribbling a note on a piece of paper and handing it to her.

She returned to her seat at the helm and begun recording the message. "This is the captain speaking. We are now cooperating with Proximian security forces aboard our ship. Currently, all non-essential crew are to remain in their quarters. All security personnel are to return their weapons to the armouries and then return to their quarters. I will remain in command and we are actively working to reestablish normal operations."

## Chapter 36

# CORNERED ANIMALS

With no way to escape, putting distance between them and the enemy about to breach the door wasn't an option. Their best choice was to barricade the end of the hallway, which meant moving the operators out. Cramping everyone into such a tight space wasn't practical, and Kane didn't want the crew in the line of fire.

The captain tasked one crew member to man the terminals, while the remaining crew were transferred to the former XO's living space. The team got to work on the barricade, with Hulk wrenching the captain's chair from its bolted base while the others dismantled the briefing room table. They threw together a makeshift wall in front of the control panels, angling Hulk's shield over the top as a rough roof. It was an eye saw. Not built to last, but it was the best defence option available. The hallway was now the fire lane where Kane expected to be exchanging blasts with the Schmeller's grunts.

The space was small so Kane removed anyone who wasn't essential. Unless there was a requirement for her presence, he returned the captain to her cabin with her children. Kane knew the odds of surviving what was coming and decided Davo was someone worth saving. So once again, Davo was assigned to the teens, along with their mother. Anticipating Davo's objections to his reasoning and avoided the argument by tasking him with keeping them safe and more importantly, under control. Under the captain's orders, the operator left at the terminal had no option but to stay. Following that, the team took their places in the remaining operators' seats. And the wait began.

With his rifle resting on his lap, Kane stretched his arms, interlocking his fingers as he extended them overhead. But the peace didn't last long. The captain poked her head out, catching their unsuspecting attention.

"Sorry. We need a toilet."

Kane raised an eyebrow. "And your suggestion is?"

She hesitated briefly, then grabbed a paper bin from the briefing room and disappeared back into her quarters.

Moments later, her head popped out again. "Sorry."

Then her daughter emerged from the room with the basket in hand. Avoiding eye contact, probably a smart choice, she quietly slipped into the meeting room doorway to dispose of the basket, then withdrew back into the captain's quarters.

"I hope this won't be a habit," Kane muttered, clearly annoyed by the teen.

The captain poked her head out again. "Sorry about that, sergeant. Is there anything you need?" She was clearly impatient, as most captains tended to be.

"Just a peaceful wait," Kane replied, ending the conversation.

Pope commented, "You couldn't pay me enough to swap places with Sparky at the moment."

"He's definitely got his hands full." Kane agreed.

Time dragged. That agonising stretch of time when you're utterly helpless, with nothing to do but keep waiting. Staring down the short hallway, they occasionally hearing the clanking sounds of movement echoing from the corridor. It was no surprise they all knew who was there.

Kane's eyes drifted to his hand, which faintly trembled, barely noticeable, yet unmistakable. In the quiet, he began noticing details he'd missed before. His gaze lingered on the dried blood smeared across his knuckles, a mottled blend of black and red, a grim reminder of Maihuni. He rubbed a fingertip over a stubborn patch lodged under his fingernail. The blood clinging like glue and stretching into a thin, sticky thread as he pulled his finger back. It was a grotesque tether to the violence that had unfolded, pulling him back to his last conversation with Maihuni. Not as a soldier, but as a friend.

In a sudden mix of disgust and panic, he rubbed his hand against his trousers, but the fabric had already absorbed more than he thought possible. The stains spread, dark and foreboding, taunting his futile attempt at cleanliness. Frustrated, Kane unclipped a water bottle from his belt and withdrew a manual labelled "Crew Management" from the shelving. He ripped out pages full of the names of countless Earthen crew members and scrubbed his hands. Some of the paper disintegrated into soggy clumps, clinging to his skin and mingling with the blood from around his nails. The stains were stubbornly permanent for now. At least the tremor had subsided, if only temporarily.

He glanced to his right and noticed Hulk sitting upright, eyes closed and mouth ajar. Just moments ago, he had been chatting with Pope. His head slowly tilted backward as he slept deeply, subconsciously adjusting to keep his hefty noggin balanced.

Deciding to let Hulk rest, Kane turned his attention to Pope, who was absently cleaning random gear, clearly bored. Kane grabbed a handful of blood-soaked paper, rolled it into a small ball and tossed it, narrowly missing Pope's right eyebrow. Pope shot up like a meerkat, assessing whether he'd somehow messed up or if Kane was just as bored as he was. Finally, with a smirk, he raised his middle finger and returned to his gear, brushing off Kane's attempt at entertainment.

The enemy would soon arrive, yet the wait was nothing short of agonising. Kane adjusted in his seat and began finding entertainment in watching the operator trying to do the work of four. An unmistakable taste of blood touched his tongue. He idly wondered when he had last brushed his teeth, but he forgot the thought when the captain's head once again appeared at the doorway.

"May I join you?" she asked, believing she had given him the time he'd requested.

"You may," Kane replied, feeling a flicker of impatience with the lull in activity. He welcomed the distraction, even if it came with added risk. If she wanted to put herself in the line of fire, that was her decision. It was her ship, after all.

The captain leaned against an operator's desk behind Kane, her voice cutting through the silence. "The soldier my daughter shot. What's his condition?"

Kane's jaw clenched at the memory. "She was bleeding out when we last saw her."

"She!" The captain's eyes widened in surprise.

"Yeah, she." Kane's mouth held the taste of blood and edged with bitterness. "She was a skilled medic, the one who held us together. The mother of our team."

The captain's face softened with regret. "I'm so sorry."

"Don't be. She knew the risks."

What he really wanted to say was, "Don't be. Your daughter's the one who shot her."

"Rayna!" Pope interjected and paused, steering the conversation in a new direction.

"Was that her name, Rayna?" the captain took the bait, her brow creasing in confusion.

"Rayna was the spy you had at the mine," Pope snapped.

The captain hesitated. "I knew we were receiving intel, but I didn't know who." A lie. She attended the meetings when her husband briefed his teams.

"Rayna was unarmed," Pope pressed on, his voice simmering with anger. "Shot point blank by one of your SERT leaders. Warrant officer fuckin' Schmeller."

The captain stiffened, taken aback by the revelation. She hadn't known of Rayna's death and the brutal circumstances of it. A heavy sigh escaped her. She hadn't expected the horrors being so personal and the weight of the mission's grim realities sank in. Listening to Pope's seething resentment, she realised he partly blamed her for the loss and she had no choice but to bear the burden of his disdain.

Kane picked up on the tension between Pope and the captain but held back, giving them space. They both needed this chance to let out their frustrations about the recent events.

She hadn't planned on sharing, but she knew she needed to shift the gears. "Those bodies in the corridor, lying in the dark," she started, her voice faltering.

"I sent them," she confessed, guilt weighing down her words.

"To save your children," Pope said, his tone more sympathetic. "But why send untrained crew against SERT?"

"My husband was the XO," she said, her gaze dropped. "It was his idea. He knew how much my children meant to me. I gave him the go ahead."

Pope said nothing but nodded slightly, concealing that he already knew this detail.

Her voice cracked as she bowed her head with the painful memories. "He died for me."

Pope's mind drifted back to the image of the teens navigating through the bodies, then the captain standing resolute, commanding her troops to stand down. The darkness had masked the faces of those on the floor, but the weight of their loss was unmistakable. He kept his silence. There was nothing to add that would benefit either of them. Her pain was clearly overwhelming. Losing her ship, many of her crew and her husband was a heavy enough burden for anyone to bear in their lifetime.

Hulk jumped awake as the operator behind him answered a call through his headset. "Command, state your emergency?"

Kane immediately recognised the operator's voice. "Not this dick again," he muttered. It was one of those moments when someone seemed normal until they opened their mouth.

The captain reached over and flicked a switch, allowing the communication to be heard on speaker.

"What do you think the emergency is, you idiot? Put the enemy commander on!" The voice bellowed.

With no emotion, the operator responded, "And who shall I say is declaring this emergency?"

"Warrant officer Hill!"

Kane couldn't help but focus on the unfazed operator. If there was anything that he and the Schmeller would agree on was that the operator was an oxygen thief.

The captain grabbed the headset from the operator. "Warrant officer Hill, this is the captain. I order you to return to your quarters."

"Captain," he paused, "With all due respect, it is my duty to protect this ship."

The captain knew her attempts were futile but tried. "I am the captain of the Tyson. We have children and crew members in here. I am working with the Proximians to establish a peaceful resolution and I order you to stand down and you will do so."

The Schmeller responded calmly, showing no concern for the captain's demands or the lives at stake. "I assume the enemy commander is listening to this. You are to lay down your weapons, unlock the door, and then lie flat on the floor. Failure to comply will result in us using force to remove the door."

The captain looked at Kane and waited for his lead.

Kane glanced at his team, knowing they shared the same opinion about facing the gunfight head-on. Much like the Schmeller, they weren't the kind to give up easily.

Kane asked the captain to relay his response.

Pope and Hulk smirked at her translation. Angered by the predicament her security team has placed her in, she added a bit of dash, "Sargeant Kane regretfully declines your offer for a friendly meet and with my full support, he has asked me to inform you to go fuck yourself."

The captain and the operator quickly made their way back to her quarters as they prepared for the blast.

Kane spat the taste of blood from his mouth and gave what he believed to be his last command before the chaos of battle. "We're like cornered animals, so let's show these fuckers what we can do."

Pope found a better analogy. "Like pigs looking for a fan to fling their shit!

# Chapter 37

# NO MIRACLE

Kane considered the Schmeller's next move. It was likely that his team would place water-based charges against the door. Likely a strip or shaped frame charge, the standard technique for breaching sliding doors like this one. Regardless of the execution method, they would soon take down the door, paving the way for a relentless onslaught of firepower directed at their makeshift barricade.

Kane communicated his expectations to the team. "We'll keep a low profile and only return fire when..."

Before he could finish his sentence, a deafening pressure wave slammed into the barricade, sending splinters flying as its supports groaned under the strain. A heartbeat later, Kane felt the bone-rattling impact. He instinctively jerked his head to the side, narrowly escaping the searing flash and a shower of debris that erupted like shrapnel. The operator terminals crackled with explosive energy. Alarms blared and smoke billowed, shrouding him in a whirlwind of chaos as his world unravelled into pure destruction.

When Kane opened his eyes, they burned from the smoke that churned through the air, its suffocating fumes carrying the rancid stench and sharp taste of burnt rubber. An eerie silence settled over the scene as he glanced at Hulk, who was still catching his breath from the dive for cover, steadying himself to strike back.

Then he noticed Pope, who was wearing a cheeky grin and shouting something. "How f— — am I? That— — it me!"

Kane could see Hulk's lips moving, but the words were lost in a chaotic blur, his hearing intermittently fading in and out. The scene felt distorted, and it took him a moment to piece together what was happening.

His other senses kicked in, helping him piece together the unfolding events. He spotted the door brace wedged halfway through the barricade and it hit him why Pope looked so amazed at somehow surviving the impact. Sounds filtered back in, muffled at first, like listening to childhood friends talking underwater. Then, suddenly, it all came rushing back. The shuffling of his team repositioning, the screams of the teens and the relentless pounding ring in his ears.

Everything felt distorted and his head throbbed. "I'm okay!" he shouted, though it seemed no one paid him any mind. The words were more self-reassuring than anything else.

Kane noticed the flickering of the terminal's controls, then saw the veil of smoke spiralling, being drawn beneath the terminal. His senses snapped back to high alert, jolting him from his dazed state. "Oh, fuck!" he thought, before shouting, "Mask! Mask! Mask!"

The team turned to follow Kane's outstretched arm, eyes fixed on the swirling smoke. A minor breach was venting their air supply into the vacuum beyond. A consequence of the enemy's excessive use of explosives that had expanded the blast zone to the ship's hull. Instinct kicked in. Rifles and helmets clattered to the floor as they tore at the Velcro packs on their belts, swiftly strapping on their emergency masks. Kane adjusted his mask until he was satisfied with the seal, then he re-secured his helmet in place. He grabbed his rifle, aiming it over the edge of the barricade, poised for whatever threat might emerge. But nothing came.

The team's focus shifted to the exposed corridor at the end of the hallway. The other doors stood battered but intact, marked only by a few scratches from the breached one, now twisted and lying in a pool of gel on the floor. Charred sections of the door frame were dangling precariously, but there was no trace of the enemy. An eerie silence filled the space, broken only by muffled whimpering coming from the captain's quarters.

Two canisters clattered into the hallway, their metallic echoes piercing the silence. Kane ducked behind the barricade, squeezing his eyes shut as the cylinders spun to a stop. A breathless moment passed, making him wonder if they were duds.

Then, twin blasts detonated with a brutal force that slammed into the barricade. The blinding flash seared through his visor, and the deafening shockwave punched the air from his lungs, forcing him to curl inward. His ears rang in relentless waves, compounded by disorientation, but he forced himself to refocus. He forced his aiming eye open, sharpened his focus, which brought the blurry figures at the doorway into focus.

"Get down! Get down!" A voice shouted.

"Let's do this, motherfuckers!" Pope returned.

At the narrow juncture where the hallway met the corridor, the enemy fired over their shields. In brief flashes, Kane could just make out the outlines of two helmets behind the bursts of muzzle fire. Limited to the two shields and the space above for aiming, the tight entry point was a tactical bottleneck. He was also aware there were more of them ready, poised to press forward and join the fight if the leading two fell.

Energy blasts thudded, sending tremors through the upturned desk. Kane felt the heat pulsing from the waves that seeped through his protective cover, making the hairs on his neck stand on end as he revelled in the intensity. His team responded with controlled bursts of fire, keeping the enemy at bay and preventing their advance. For now, the scattered exchange down the hallway bought them precious time, but Kane knew it wouldn't last. Squinting against the barrage of colourful blasts ricocheting off the angled shield above, he noted the stalemate and ordered his team to conserve their battery power.

The XO's and captain's quarters remained sealed and safe from the vacuum that was slowly draining oxygen from the surrounding areas. Kane realised the enemy hadn't noticed the leak yet as they weren't reaching for their masks. But once they did, the air itself would become their deadliest threat. It seemed they, too, had recognised the stalemate and had slowed their rate of fire.

A shot ricocheted off Pope's visor, grazing his temple and leaving a shallow gash. His legs buckled from the blow, blood smearing across his visor as he tried to wipe it away with a gloved hand. Though the wound wasn't fatal, it was disorienting. He ripped off his mask and fumbled for a two field dressings. One to clear his vision as the other patched the wound. Meanwhile, Kane and Hulk kept up a steady return fire, knowing the enemy wouldn't let up.

As the gunfire lessened, flashes of deadly shots still lit up the hall. "What are they up to?" Hulk muttered, echoing Kane's thoughts. A chill crept over Kane, not just from the warmth draining through the damaged hull, but from a cold sense of dread gnawing at him. He knew the situation wouldn't get any better. Time was slipping through their fingers, and every second was becoming critical.

Once the enemy noticed the leak, they could retreat to safer ground, forcing Kane's team to seek refuge in the self-contained crew quarters. Those quarters would become their prison and soon, they'd be gasping for air. Kane knew he had no choice. He had to come up with a new plan, and fast. The only option he saw was to rush the enemy. The odds of survival were slim, and he and his team would likely die in the attempt, but with suffocation as the alternative, they were forced to take the gamble.

Kane theorised as Hulk nudged him. "Check out behind the one on the right."

Peering through a gap, Kane spotted the figure through the enemy's perspex shields. "The Schmeller," he murmured, recognising him immediately.

The dynamics shifted in an instant, which seemed to be the norm when the Schmeller was present. To Kane's horror, he caught sight of the PEPS (Pulsed Energy Projectile System) balanced on the Schmeller's shoulder. Bracing himself, the Schmeller rose behind the shields, struggling to angle the heavy weapon into a horizontal firing position within the cramped corridor. Kane knew their stand-off was about to reach a brutal end.

The system unleashes a concentrated pulse of blinding plasma that crackles with raw energy. The searing blast would strike the command centre, erupting outward in a violent wave, engulfing the area in a cascade of energised fury. In the tight confines at the end of the narrow hall, a single, devastating shot would wipe out Whiskey. If carrying such a cumbersome weapon had been practical, Kane would have done the same.

Kane was the first to react, knowing the others would have done the same. Without hesitation, he sprang to his feet, urgency fueling every movement. He knew any delay could be fatal. His hand shot up, yanking the shield free from its makeshift mount. It swung forward, miraculously angling just right, held steady by the door brace wedged into the desk. Kane had the perfect clearance. His eye locked onto the Schmeller and without wasting a moment, he squeezed the trigger. His firing blast shot forward before his eye aligned with his sights.

Thump—Thump—Thump. The cadence echoed down the hallway, unleashing a relentless barrage of fire, interrupted only by Kane's raw, furious scream—a primal roar of defiance. The enemy, crouched behind their shields, hesitated, briefly thrown off by Schmeller's movement. But Kane's adrenaline-fueled onslaught snapped their focus back.

Shot—Recharge—Shot—Recharge. The relentless rhythm allowed them no chance to respond. With single-minded focus, Kane squeezed the trigger, firing again and again, undeterred until he found his mark.

The initial blast struck their shield wall with a heavy thud, sending electric tendrils skittering across its surface. The next few shots climbed steadily, sizzling up the shield until they reached the edge. One bolt veered wide as Kane zeroed in, but the next was the money shot. It struck the Schmeller's leading hand, severing two fingers and scorching a third. The impact jolted the PEP weapon upwards, just as the Schmeller squeezed a reactive shot in return.

Kane barely had time to react and dove for cover. In that split second, he shifted from imminent doom to becoming the dealer of death. A surge of electromagnetic chaos rippled through the command centre, crackling along the ceiling and walls. Manuals, stationery and shelves flew like confetti. Whiskey felt the shockwave slam against their overturned barricade, screeching as it twisted and pinned them against the desks. The blast yanking their masks tight against their faces.

The three of them trembled, their bodies aching from the disorientation. Kane took a deep breath, grounding himself in the realisation that he was still alive. It wasn't a miracle that saved Whiskey, unless you counted a resolute leader and a perfectly aimed strike as one. Kneeling, Kane fought through the haze, shouting urgently for Pope and Hulk to regroup. They had to move fast, or their brief reprieve could spiral into catastrophe.

"We can't stay here!" he barked. "We need to move now!"

# Chapter 38

# BLINKING EYES

Whiskey regained focus as they pushed against the barricade. As the ringing in their ears faded, the alarms grew louder, heightening the tension of their predicament. Aside from Pope's helmetless, bandaged head, both he and Hulk had come through relatively unscathed. Steadying their weapons, they moved into the battle scared hallway, driven by their leader's unwavering resolve to face what the enemy dealt next.

Hulk retrieved his battle-scarred shield, knelt in front of the shattered remains of the desk, and waited in silence. The narrow view of the corridor was filled with the twisted limbs of bodies, lying motionless in a haze of smoke. With no immediate threat in sight, they pulled themselves together, adjusting their masks, swapping out depleted batteries, and gathering their thoughts.

Kane took his place between Hulk and Pope, casting a quick look at the wrecked control panels. "They won't be givin' orders from here anytime soon." He said.

Pope glanced over but, for once, said nothing. The toll of battle was weighing on him.

Kane recognised the exhaustion etched on Pope's face, a weariness he knew all too well. Each surge of adrenaline drained their reserves a little more, the energy that had carried them through the firefights now fading, replaced by a heaviness that made every movement and decision feel like a struggle. Pope saw it too, and after a moment, his battle-hardened expression returned. He gave Kane a nod, and together they fell back into formation.

Hearing a crackle in his headset, Kane tried establishing coms with the Centaurus. Nothing. "Coms are down," he murmured.

He tapped Hulk on the shoulder, who hauled himself to his feet.

"Advance," Kane whispered. Hulk raised his shield and began moving forward.

Kane's pulse quickened as his instincts kicked in at the faint scrape of metal in the distance. Whatever lurked beyond the smoky corner wasn't friendly. They advanced slowly, tense and silent. A low whistle pierced the air as the small vortex squeezed through the hull breach behind them. Not life-threatening yet, but the temperature was dropping fast.

As they approached the shattered doorway at the end of the corridor, the scraping of the looming threat grew louder, its urgency thickened by laboured breathing. Kane tapped Hulk's shoulder, signalling the team to halt. He slipped silently past the shield, crouching low, inching closer to the hallway's edge. His heart pounded as he stole a quick, perilous peek at whatever was approaching.

His glance into the darkness revealed a grim scene of bodies scattered along the corridor. Those close were motionless, their bodies twisted in unnatural angles. Those furthest from the blast were scattered in an eerie formation, their bodies splayed out like the quills of a porcupine. A few still moved, weakly clawing at the floor, their fingers digging into others as if grasping for any last shred of life. None were upright, nor had the strength to stand.

The source of the threatening sound was a lone soldier, his body burnt and broken, dragging himself slowly toward the doorway. His body armour scrapping against the floor, his kevlar armour grated on the metal with each agonising forward pull. His legs refused to cooperate, limp and lifeless behind him, yet he kept moving, his trembling hands reaching out for help. The soldier grew more desperate, a low gurgle escaping his lips as his strength slowly wilted. Kane's heart sank at the sight. His mind drifted back to Maihuni, her twisted form lying on the floor. Now, all he wished was that he had allowed her to be more a part of his life.

Unsure who, or what, may still lurk, Pope and Hulk advanced cautiously past the bodies. Each searching the opposite ends of the corridor to ensure there were no hidden threats waiting for them. There were none.

Kane lifted his mask, testing the oxygen levels but the stench found his nostrils. He quickly refitted his mask. He found the problem. A damaged pipe was bubbling and hissing. An enemy shield had ruptured the lining of a holding tank within the wall. The pipe oozed and gurgled with a mixture of urine, human waste and partially dissolved paper. Far from being drinkable, he guessed the odour wouldn't even be suitable for irrigation.

Kane gathered the masks from those who no longer needed them and pounded on the XO's door. "It's Kane! Open up!"

As the door slid open, the occupants gasped, a primal instinct gripping them, fearing it might be their last breath. The vacuum of the hallway instantly sucked the precious air from their small sanctuary, replacing it with the stench.

"Put these on." Kane handed out the masks he had collected. He gestured to the compromised section of the wall. "We've got a leak." Their noses made it clear it wasn't the only leak that needed fixing.

The crew scrambled and searched for the emergency repair kit, which had been dislodged somewhere in the command centre.

Kane repeated the events at the captain's door. He removed both Davo and the captain. Her children stayed behind for their safety. The captain read between the lines, understanding that their fragile nature would make it difficult for them to comprehend the carnage in the corridor.

"Davo, I need you and Hulk to escort the captain to the command centre in the forward superstructure," Kane ordered.

"On it." Davo waited patiently as the captain gathered her essentials.

"And captain," Kane added, "once you establish contact with your crew, get medical and repair teams down here ASAP."

Turning back to Davo, he added, "Contact our ship. Give them an update and request a security team to take command."

Kane returned to the corridor, where his team tended to the wounded. "Hulk, go with Davo when they head out."

"Roger that," Hulk replied.

As Kane surveyed the corridor, with its walls splattered like a Pro Hart masterpiece, he reflected on the odds of his survival. It wasn't because his cause was righteous, his enemies had believed theirs was too. Now many of them lay dead and he was the one still standing. The victor.

To his surprise, Kane noticed a flicker of movement. Blinking eyes amongst the charred bodies at the base of the wall.

Warrant officer fuckin' Schmeller. Once the very embodiment of arrogance, strutting out of the lift as if he were invincible. Now, his body betrayed him, no longer matching the untouchable ego he once wore like armour. Now, with his bloodied eyes staring blankly past his shattered nose.

"Well, well," Kane muttered. The Schmeller's eyes focused towards him. "Looks like the Schmeller is harder to kill than I thought," he said to Pope.

"Show him it's not as hard as he thinks," Pope recommended coldly.

Kane felt no remorse. This failure of a man went against doing what was right, both legally and morally. If he had done the right thing, Rayna would still be alive. Kane considered compromising his own standards to match the Schmellers. However, the universe didn't need another man like him.

The Schmeller's gaze flickered with desperation as he attempted to plead, but his bloodied throat only managed a gurgling choke. Kane recognised the fear in his enemy and hesitated, questioning whether he should sink to the level of the man he despised.

"May I?" Davo asked quietly.

Kane stepped aside, watching as Davo raised his SIPLAT, aiming it at the cornered man. His face remained emotionless. No hint of satisfaction in avenging Rayna and no pleasure in the thought of ending the life of the evil in his universe.

"Just know," Kane said, "this moment will most likely be a lifetime of recurring nightmares."

"A fair trade," Davo replied coldly, his focus locked on the man before him.

The Schmeller, resigned to his fate, closed his eyes and without hesitation, Davo pulled the trigger. Nothing happened. His cold stare shifted to confusion as he realised his act of vengeance had failed.

"Safety," Kane reminded him.

Davo flipped the safety off and aimed again. This time he hesitated. He had never fired a weapon before, not this one or any other. His first shot would end a life. His hand trembled slightly, but most likely aided by the growing chill in the corridor. The others watched in silence, none of them willing to intervene.

The image of Rayna flashed in his mind and with a held breath, he pulled the trigger.

A hole erupted through the Schmeller temple and he exhaled his last blood bubbling breath. Splattering matter oozed out of the back of his skull. He became nothing more than a haunting memory.

Davo heard someone nearby murmur, "Welcome to the team," but he remained frozen, staring blankly.

"Look at me," Kane said firmly, shaking Davo to break his focus. "You need to get the captain out so we can get the wounded some help."

Davo snapped back to reality. Satisfied with his desire for vengeance, he moved on, as did Kane.

The captain's voice crackled over the coms once she reached the forward superstructure, rallying repair teams with swift, steady orders. Responders in grim focus moved quickly to assess and patch the damage, as medical teams finally relieved Pope and Kane. Escaping the blood-soaked corridors and the quiet suffering of the wounded, they retreated to the captain's meeting room. The air was easier to breathe here and the stench of waste and death was less overpowering.

Kane slumped into the captain's empty swivel chair, his head tilted back, eyes tracing the cold pipes and beams above. His eyes unfocused as if he were miles away.

After a moment, he broke the silence with a muttered confession, simple yet loaded. "I'm knackered."

Pope stared towards the door, expression unreadable. "Yep."

"You think they still see that scratchie as a godsend?" Kane muttered.

Pope let out a dry laugh. "Godsend. Funny word, isn't it? Seems like whenever someone sends god, it's followed by a trail of death."

Kane shook his head. "Crazy, really. All this is over one little ticket. One shot at wealth, and here we are, people fighting for their lives a universe away."

"The riches are for the ones playing the political game," Pope replied, voice thick with irony. "For the rest of us, just dirty work, long hours and lost friends."

Kane ran a hand over his face, eyes still on the beams above. "Makes you wonder why we haven't learnt that the people who pick the fight ought to be the ones fighting it."

Pope glanced over, a glint of humour in his tired eyes. "If that ever happened, we'd both be out of a job."

Kane met Pope's eyes. "So, what do you reckon our prize is then?"

"Maybe it's just that we're still here to ask that question."

As the panicked commotion outside faded, they slipped out of the meeting room. The crew members had completed temporary repairs, with a few still clearing the fallen. Kane heard the teens calling out for attention. He shouted, "Stay inside!" He left them safely where they were, away from the choking stench of sewage, death and the possible potential wrath of Pope.

Once the situation had stabilised, Kane handed over command to the Centaurus' XO, who arrived with the remnants of Papa and Quebec. Together, they moved methodically through the ship's corridors, leaving no corner unchecked. The XO made it clear: until the critical repairs were finished and orders arrived from Proxima, the Tyson wasn't going anywhere.

Chapter 39

# WE'D BOTH BE WRONG

With Whiskey's mission complete, they returned to the Centaurus, embracing the simple comforts of hot water and fresh clothes. It wasn't long before life returned to its usual rhythm as the situation around the Tyson came under control.

After a few clandestine swigs of mead in his quarters, Kane made his way to the amenities. Moving down the corridor, he found himself once again stopping at the portal, alone with his thoughts, staring out into the vast, indifferent expanse. His gaze drifted as he watched the tenders carefully guide the BioStat back into position, where it began its final approach. The delicate operation edged the tin can closer. Shortly, its locking mechanisms will unite it with Jingai.

He found it surreal when every object locked into the same velocity and vector. The illusion of stillness took over. Despite hurtling through the vastness of space at immense speed, everything appeared perfectly motionless, as if the universe itself were inert.

What began as a gamble on a small scratchie had spiralled into a bloody conflict, claiming the lives of countless people. To most, they were nameless, faceless figures lost in a soon to be forgotten battle. He knew it was one that he would never forget.

The misguided pocket rocket sized daughter, of a treasonous father. Kane allowed his thoughts to wander to Rayna and what would never be. He knew the memories of her would now remain a part of him and he found solace in knowing he was a better man for the short time they'd had together.

Kane knew his haunted memories would shadow him wherever he went, and the thought of changing careers had come and gone. He accepted that his place was with SERT, where he was surrounded by others who understood him. The ones who shared the same night sweats. Here, Kane was amongst those fighting demons, much like his own.

The engineer, whose quarters were close by, passed Kane on his way to the amenities.

"Has your grandson arrived yet?" Kane asked, recalling their last conversation.

The engineer stopped, his face breaking into a wide grin. He extended a hand in gratitude for Kane's thoughtfulness. "A granddaughter, actually."

"Congratulations," Kane replied warmly, clasping the engineer's hand. "What's her name?"

"Maihuni," the engineer said, pride clear in his voice.

Kane raised an eyebrow, pleasantly surprised.

The engineer smiled. "I told my daughter about your team's miraculous adventure. She loved the story of Maihuni and named the baby after her."

Kane was momentarily speechless. He'd never thought of his team as anything more than the XO's problem-solvers. People in his line of work rarely ended up as heroes in anyone's story.

"Well," he replied, feeling a little out of his element, "I can say that the baby is definitely named after one tough woman." It wasn't something he'd typically say and sounded almost pretentious, but it felt right in the moment. The engineer, absorbed in his joy, nodded and continued on his way, leaving Kane to process the unexpected honour.

Kane turned and made his way to the Ring, hoping to slip past the upcoming presentations with little fanfare. But the victory celebrations were far from winding down. The captain's invitation had attracted a large crowd, with tables reserved for department heads and those directly involved in the off-ship missions, leaving standing room only for many others. The Ring was rarely this packed, especially with the added presence of guests from BioStat.

In his line of work, recognition was more a nuisance than a reward. Kane preferred to concentrate on doing his job well, giving his best for his team, and repeating that same commitment every shift. He knew being acknowledged for excellence only bred arrogance, and he recalled all too well how that had played out for Schmeller.

Kane approached the booth marked with a handwritten sign reading "Reserved - SERT Whiskey."

"How's it going, boss?" Hulk greeted him as he settled in with the team.

"I don't feel the love for these shindigs," Kane replied, rubbing his temple.

Pope handed him a glass. "A few shots of eye opener will fix that."

Kane glanced at Maihuni. "So, how's the recovery coming along?"

"I'm hanging in there," she said, though the strain of her recovery was clear. "Not sure how much longer I can stay here, but the captain wanted me to be here with the team.

"We wanted you here with the team. The captain just made it happen." Kane clarified.

A warm smile spread across Maihuni's face as she felt the sense of belonging that came with being part of this family.

"I wonder if that scratchie prize the Earthlings were after still holds the same importance for them now," Pope mused, taking a sip from his own glass.

"Hopefully they will go play somewhere else." Hulk gave his thoughts.

"Only time will tell." Kane wasn't bothered with events that were way above his pay grade.

Davo entered, dressed to impress and clearly pleased with himself. Kane shuffled over as Pope poured a shot for Davo and made sure everyone else's glasses were filled to high tide.

As the voice of reason, Kane reminded, "Let's not down too many until the formalities are over."

"Got it," Pope replied, setting the bottle beside him.

"Well, aren't we the sparky one?" Maihuni remarked, unintentionally drawing attention to Davo's eye-catching outfit.

"So, it's Sparky Sparky now, huh?" Hulk added, eager to keep the banter going.

"Only if you've got a stutter, you dickhead," Kane replied with a grin.

They raised their glasses, the mead burning as it slid down their throats. Davo let out an enthusiastic "Ahhh!" as he felt the heat hit. "I've never felt more alive," he declared, his voice rasping.

Kane felt a wave of relief seeing that Davo appeared to have a clear conscience. Maybe the chaos they left behind could eventually be tucked away in their memories.

"So, where's your rabbit?" Maihuni asked Hulk.

"It should be here with us, but I got a hard 'NO.' Apparently, we're all equal, except for our mascot."

"Don't be ridiculous, Hulk. Our mascot's a wild pig, not some fluffy rabbit," Maihuni said.

Pope grinned at her. "Actually, you missed it. We court-martialled the rabbit on two counts of misconduct."

An amused smile tugged at the corners of Maihuni's mouth as she listened.

"Miss Piggy," Pope began.

"Miss Piggy?" Maihuni repeated, her eyes wide with surprise.

"Yeah, turns out the fur ball's a 'she'. So we named her after our real mascot, the pig." Pope flashed a grin. "Anyway, Miss Piggy was court-martialed for treason after ratting out our positions to the Schmeller," he let the humor sink in, "and then we hit her with a secondary charge of... sexual misconduct."

Maihuni's head snapped up, her eyebrows shooting toward her hairline.

Hulk couldn't help but laugh. "She was pregnant, so these arseholes sentenced her to life in her hutch."

They all burst into laughter at the absurdity. Even Kane chuckling as the story played out again.

The captain entered and the room fell silent as everyone instinctively gave them their attention.

"I want to begin by recognising the outstanding efforts of our security teams," the captain said. "They've taken on challenges most of us consider too dangerous." He paused, his tone sombre. "And as you all know, we've lost lives defending what is ours."

He continued, "In one cycle, at three bells, we'll honour our fallen. The service will be streamed to Proxima, so I expect everyone in their finest dress."

His gaze swept across the room. "I'm confident the fallen would want us gathered to thank each of you for your brave efforts in defeating a worthy adversary."

"To the fallen!" someone called, raising a glass. The room echoed the salute, glasses lifted high. "To the fallen!"

As the voices quietened, the captain added, "Before we celebrate, I have some updates. You may have seen the news. They arrested the father of the spy on the BioStat for treason.

The room was silent as the captain continued his briefing. "As for the attempted assault by Earth, you'll be relieved to hear there are no other Earthen ships approaching Jingai."

From interrogations aboard the Tyson, intel suggests Earth isn't as prosperous as they like to claim. They're resorting to desperate measures in their push to expand into space, even though they have no legal claim to what's ours. We've halted their attempts to resolve this with force, and the matter is now in the hands of the courts. So, until this is settled, we are officially at war with Earth."

He double-checked his notes before offering his opinion. "As for Earth, I have just one thing to say." He paused, then grinned. "Fuck 'em!"

Everyone started laughing and raised their glasses. "Fuck 'em!" echoed back.

The captain raised his voice, silencing those gathered. "As you all know, the XO and a security team are aboard the Tyson. It's undergoing repairs, and its future remains uncertain. I'm guessing it'll be released back to Earth as part of a larger deal. Until then, he's asked me to stand in for him as we honour our brave."

Shifting into a more solemn demeanour, he continued, "On behalf of Proxima, the crew of the Centaurus and our distinguished guests from Jingai, I'd like to call Sergeant William Kane to the front."

A murmur rippled through the crowd as Kane moved toward the captain. There was an odd air of uncertainty among the onlookers. The captain was obviously unaware of the mystery surrounding Kane's given name. Most knew him simply as Kane, unaware it was his surname, and the truth behind his name was now revealed, adding a layer of intrigue to the moment.

With Proxima's approval, SERT members involved in the Tyson incident received citations for their actions, and Kane accepted the honour on their behalf. The XO requested a low-key ceremony, understanding the team would prefer to avoid a public celebration. They were, after all, just doing their job.

Then Davo was called forward, celebrated like a hero. The crowd, if not wanting, seemed to need a hero. Unlike Whiskey, Davo lapped up the recognition and the people embraced him for it.

The captain used the opportunity to further unite his crew and praised Davo's selfless courage in risking his life for them. "Not only does he receive this award," he announced, "but the captain of the Tyson also gave him command of her ship, though his command was brief."

The crew raised their glasses. "To Captain Davo!"

With the formalities over, the celebrations kicked into high gear.

Maihuni seized the moment. "I don't know if it's the drinks mixing with the antibiotics, but I'm feeling dizzy. I need to hit the fart sack."

Kane offered his arm, helping her back to her quarters. As he settled her onto the bed, the moment presented itself for them to speak openly, the unspoken tension between them palpable.

Kane had long suppressed his feelings for Maihuni. After all, she was one of his soldiers and that was a line he never thought he'd cross. But after she was shot, he realised he couldn't bear the thought of losing what he had with her, as he had with Rayna.

They both sensed where this was heading.

"I'm your boss," Kane said, his tone firm but edged with uncertainty. "We shouldn't go there, Huni."

She looked him straight in the eye. "You know, if I agreed with you, we'd both be wrong."

With that, the line between them blurred, and their connection deepened beyond anything Kane had expected. Once again,

Kane found himself caught between duty and desire, mixing business with pleasure in a way that felt both reckless and irresistible. This night wasn't just a fleeting moment. It would be one of many chapters in the ongoing story of Kane and Maihuni, a story that neither of them wanted to end.

# ABOUT THE AUTHOR

Allan's quest for the perfect story spanned two years, combing through libraries and bookstores in search of a novel that aligned with his passions. When that search proved fruitless, he took matters into his own hands, igniting a journey of writing that rekindled a youthful thirst for creativity. His home art studio became the hub of this transformation, where casual gatherings with friends, once focused on art, shifted to plot brainstorming. This gave way paintbrushes for whiteboards and mind maps. What began as a social ritual soon became fertile ground for his storytelling.

Allan sought to combine his military knowledge with science fiction, aiming to maintain a sense of realism within an imagined or hypothetical future. Having served with the 5th/7th Battalion, Royal Australian Regiment in his twenties, the skills and experiences Allan gained during those years are what he strives to share in his writing. The old saying, "Once a soldier, always a soldier," has defined much of his life.

After his military service, Allan embarked on a twenty-year career as an educator, teaching english, history and geography. His creativity didn't fully blossom until after his teaching career, when he opened an art studio. There, he shared his creative skills with others while crafting his own works of art. This venture sparked nights of creativity with friends and marked the beginning of his writing journey.